Advanced Rigger

Trainee Guide

Boston   Columbus   Indianapolis   New York   San Francisco   Amsterdam
Cape Town Dubai London Madrid Milan Munich Paris Montreal Toronto Delhi
Mexico City Sao Paulo Sydney Hong Kong Seoul Singapore Taipei Tokyo

NCCER

President: Don Whyte
Vice President: Steve Greene
Chief Operations Officer: Katrina Kersch
Rigger Curriculum Project Manager: Chris Wilson
Senior Development Manager: Mark Thomas

Senior Production Manager: Tim Davis
Quality Assurance Coordinator: Karyn Payne
Desktop Publishing Coordinator: James McKay
Permissions Specialists: Kelly Sadler
Production Specialist: Kelly Sadler
Editors: Graham Hack, Debie Hicks

Writing and development services provided by Topaz Publications, Liverpool, NY

Lead Writer/Project Manager: Troy Staton
Desktop Publisher: Joanne Hart
Art Director: Alison Richmond

Permissions Editor: Andrea LaBarge
Writers: Troy Staton, Thomas Burke, Terry Egolf

Pearson

Director of Alliance/Partnership Management: Andrew Taylor
Editorial Assistant: Collin LaMothe
Program Manager: Alexandrina B. Wolf
Assistant Content Producer: Alma Dabral
Digital Content Producer: Jose Carchi
Director of Marketing: Leigh Ann Simms

Senior Marketing Manager: Brian Hoehl
Composition: NCCER
Printer/Binder: LSC Communications
Cover Printer: LSC Communications
Text Fonts: Palatino and Univers

Credits and acknowledgments for content borrowed from other sources and reproduced, with permission, in this textbook appear at the end of each module.

ISBN-13:   978-0-13-518321-2
ISBN-10:   0-13-518321-9

ScoutAutomatedPrintCode

Preface

To the Trainee

Rigging is an activity performed on virtually every construction site and in every industrial facility. Becoming a trained rigger will open up the doors of opportunity and you will be able to choose to work in construction, the power industry, the petroleum industry, the maritime industry, the mining industry, or the manufacturing industry.

As you progress through your training, you will be advancing your knowledge and skills with progressively challenging topics and activities. Riggers who continue to advance their knowledge have plenty of room for job growth. After gaining experience, advanced riggers can become rigging foremen or lift directors.

The need for riggers will continue to increase in all industries as experienced riggers retire. Those who meet the qualifications under OSHA 29 *CFR* Part 1926 Subpart CC, *Cranes and Derricks in Construction* will continue to be in high demand in this field that the US Department of Labor expects to experience faster than average growth over the next several years.

New with *Advanced Rigger*

NCCER is pleased to release *Advanced Rigger* in an updated format, with new photographs and figures. This version is now presented in NCCER's improved instructional systems design, in which the sections of each module are directly tied to learning objectives.

The training in *Advanced Rigger* includes an added 30 hours of training. The "Load Charts" module offers a thorough review of how to use crane load charts for lift planning. It also offers additional examples and study questions on load calculations. The "Lift Planning" module covers the most important sections of ASME P30.1, *Planning for Load Handling Activities*. The "Hoisting Personnel" module is greatly expanded to include the standards from ASME Standard B30.23, *Personnel Lifting Systems*, and all regulations from OSHA 1926.1431, *Hoisting Personnel*.

We wish you success as you progress through this training program. If you have any comments on how NCCER might improve upon this textbook, please complete the User Update form located at the back of each module and send it to us. We will always consider and respond to input from our customers.

We invite you to visit the NCCER website at **www.nccer.org** for information on the latest product releases and training, as well as online versions of the *Cornerstone* magazine and Pearson's NCCER product catalog.

Your feedback is welcome. You may email your comments to **curriculum@nccer.org** or send general comments and inquiries to **info@nccer.org**.

NCCER Standardized Curricula

NCCER is a not-for-profit 501(c)(3) education foundation established in 1996 by the world's largest and most progressive construction companies and national construction associations. It was founded to address the severe workforce shortage facing the industry and to develop a standardized training process and curricula. Today, NCCER is supported by hundreds of leading construction and maintenance companies, manufacturers, and national associations. The NCCER Standardized Curricula was developed by NCCER in partnership with Pearson, the world's largest educational publisher.

Some features of the NCCER Standardized Curricula are as follows:

- An industry-proven record of success
- Curricula developed by the industry, for the industry
- National standardization providing portability of learned job skills and educational credits
- Compliance with the Office of Apprenticeship requirements for related classroom training (*CFR* 29:29)
- Well-illustrated, up-to-date, and practical information

NCCER also maintains the NCCER Registry, which provides transcripts, certificates, and wallet cards to individuals who have successfully completed a level of training within a craft in NCCER's Curricula. *Training programs must be delivered by an NCCER Accredited Training Sponsor in order to receive these credentials.*

Special Features

In an effort to provide a comprehensive and user-friendly training resource, this curriculum showcases several informative features. Whether you are a visual or hands-on learner, these features are intended to enhance your knowledge of the construction industry as you progress in your training. Some of the features you may find in the curriculum are explained below.

Introduction

This introductory page, found at the beginning of each module, lists the module Objectives, Performance Tasks, and Trade Terms. The Objectives list the knowledge you will acquire after successfully completing the module. The Performance Tasks give you an opportunity to apply your knowledge to real-world tasks. The Trade Terms are industry-specific vocabulary that you will learn as you study this module.

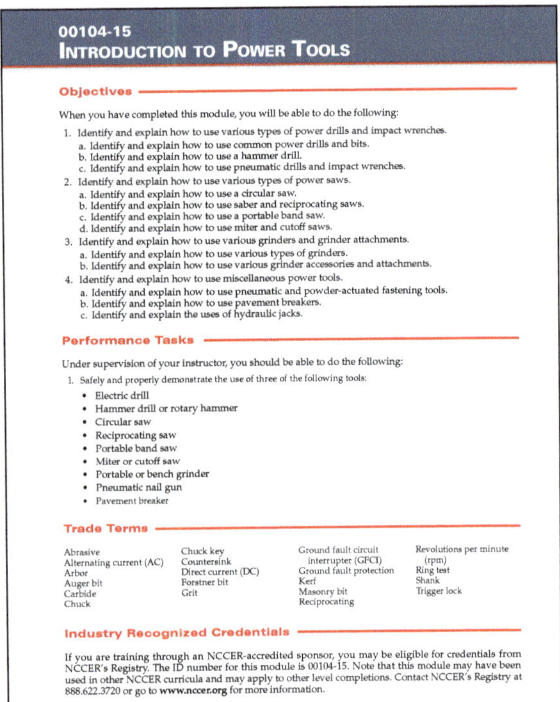

Trade Features

Trade features present technical tips and professional practices based on real-life scenarios similar to those you might encounter on the job site.

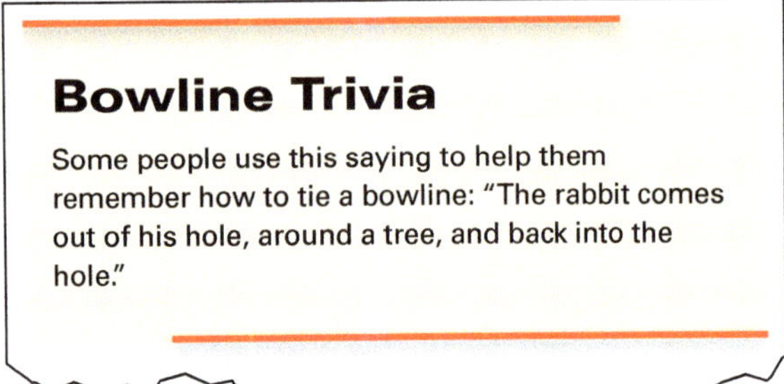

Figures and Tables

Photographs, drawings, diagrams, and tables are used throughout each module to illustrate important concepts and provide clarity for complex instructions. Text references to figures and tables are emphasized with *italic* type.

Notes, Cautions, and Warnings

Safety features are set off from the main text in highlighted boxes and categorized according to the potential danger involved. Notes simply provide additional information. Cautions flag a hazardous issue that could cause damage to materials or equipment. Warnings stress a potentially dangerous situation that could result in injury or death to workers.

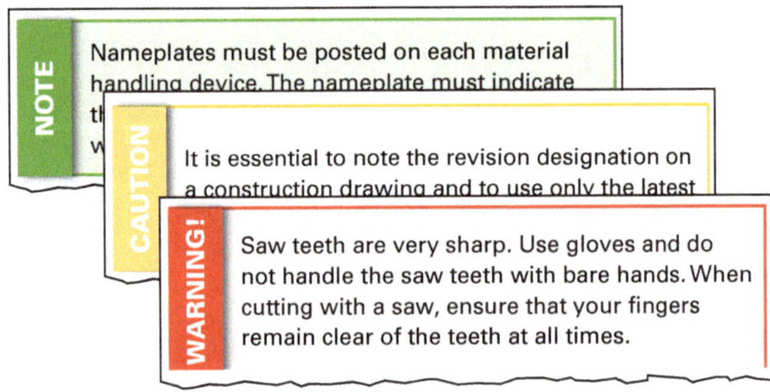

Case History

Case History features emphasize the importance of safety by citing examples of the costly (and often devastating) consequences of ignoring best practices or OSHA regulations.

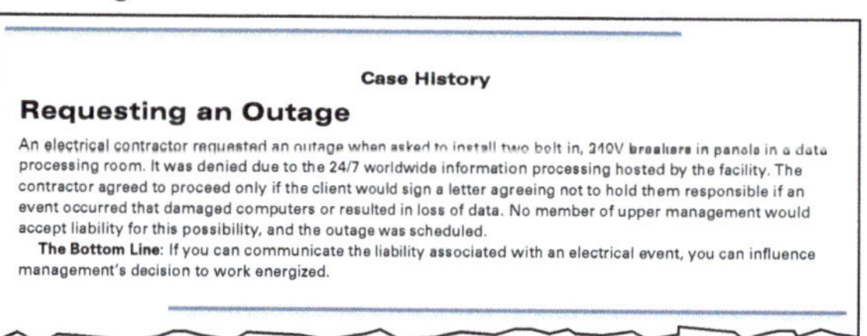

Going Green

Going Green features present steps being taken within the construction industry to protect the environment and save energy, emphasizing choices that can be made on the job to preserve the health of the planet.

Did You Know

Did You Know features introduce historical tidbits or interesting and sometimes surprising facts about the trade.

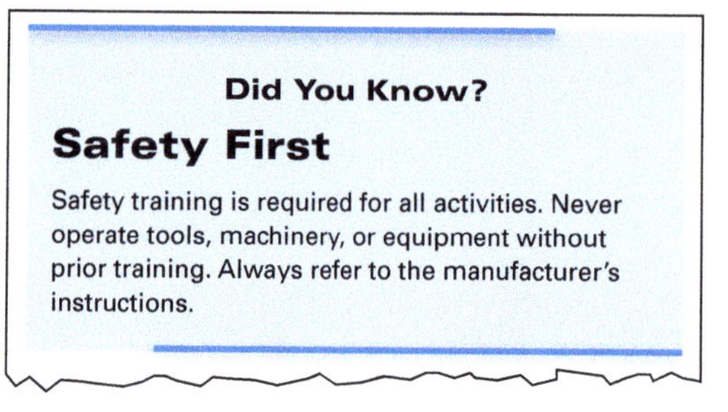

Step-by-Step Instructions

Step-by-step instructions are used throughout to guide you through technical procedures and tasks from start to finish. These steps show you how to perform a task safely and efficiently.

Perform the following steps to erect this system area scaffold:

Step 1　Gather and inspect all scaffold equipment for the scaffold arrangement.

Step 2　Place appropriate mudsills in their approximate locations.

Step 3　Attach the screw jacks to the mudsills.

Trade Terms

Each module presents a list of Trade Terms that are discussed within the text and defined in the Glossary at the end of the module. These terms are presented in the text with **bold, blue** type upon their first occurrence. To make searches for key information easier, a comprehensive Glossary of Trade Terms from all modules is located at the back of this book.

During a rigging operation, the **load** being lifted or moved must be connected to the apparatus, such as a crane, that will provide the power for movement. The connector—the link between the load and the apparatus—is often a sling made of synthetic, chain, or **wire rope** materials. This section focuses on three types of slings:

Section Review

Each section of the module wraps up with a list of Additional Resources for further study and Section Review questions designed to test your knowledge of the Objectives for that section.

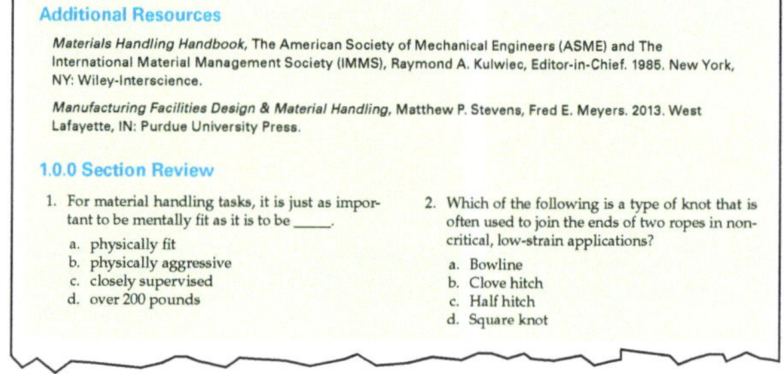

Review Questions

The end-of-module Review Questions can be used to measure and reinforce your knowledge of the module's content.

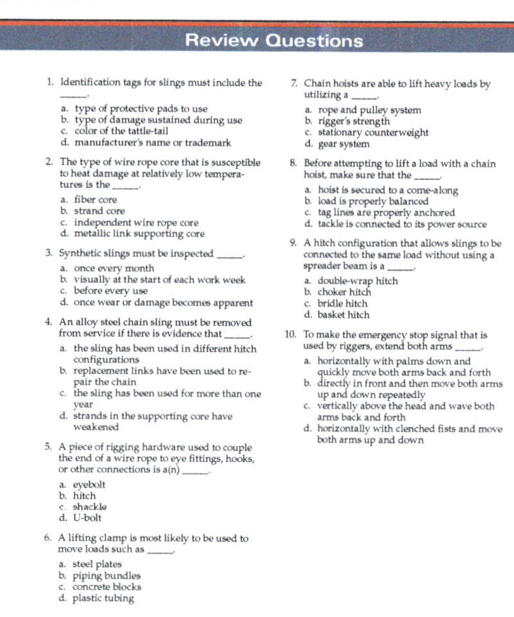

NCCER Standardized Curricula

NCCER's training programs comprise more than 80 construction, maintenance, pipeline, and utility areas and include skills assessments, safety training, and management education.

Boilermaking
Cabinetmaking
Carpentry
Concrete Finishing
Construction Craft Laborer
Construction Technology
Core Curriculum: Introductory Craft Skills
Drywall
Electrical
Electronic Systems Technician
Heating, Ventilating, and Air Conditioning
Heavy Equipment Operations
Heavy Highway Construction
Hydroblasting
Industrial Coating and Lining Application Specialist
Industrial Maintenance Electrical and Instrumentation Technician
Industrial Maintenance Mechanic
Instrumentation
Ironworking
Manufactured Construction Technology
Masonry
Mechanical Insulating
Millwright
Mobile Crane Operations
Painting
Painting, Industrial
Pipefitting
Pipelayer
Plumbing
Reinforcing Ironwork
Rigging
Scaffolding
Sheet Metal
Signal Person
Site Layout
Sprinkler Fitting
Tower Crane Operator
Welding

Maritime

Maritime Industry Fundamentals
Maritime Pipefitting
Maritime Structural Fitter

Green/Sustainable Construction

Building Auditor
Fundamentals of Weatherization
Introduction to Weatherization
Sustainable Construction Supervisor
Weatherization Crew Chief
Weatherization Technician
Your Role in the Green Environment

Energy

Alternative Energy
Introduction to the Power Industry
Introduction to Solar Photovoltaics
Power Generation Maintenance Electrician
Power Generation I&C Maintenance Technician
Power Generation Maintenance Mechanic
Power Line Worker
Power Line Worker: Distribution
Power Line Worker: Substation
Power Line Worker: Transmission
Solar Photovoltaic Systems Installer
Wind Energy
Wind Turbine Maintenance Technician

Pipeline

Abnormal Operating Conditions, Control Center
Abnormal Operating Conditions, Field and Gas
Corrosion Control
Electrical and Instrumentation
Field and Control Center Operations
Introduction to the Pipeline Industry
Maintenance
Mechanical

Safety

Field Safety
Safety Orientation
Safety Technology

Supplemental Titles

Applied Construction Math
Tools for Success

Management

Construction Workforce Development Professional
Fundamentals of Crew Leadership
Mentoring for Craft Professionals
Project Management
Project Supervision

Spanish Titles

Acabado de concreto: nivel uno (*Concrete Finishing Level One*)
Aislamiento: nivel uno (*Insulating Level One*)
Albañilería: nivel uno (*Masonry Level One*)
Andamios (*Scaffolding*)
Carpintería: Formas para carpintería, nivel tres (*Carpentry: Carpentry Forms, Level Three*)
Currículo básico: habilidades introductorias del oficio (*Core Curriculum: Introductory Craft Skills*)
Electricidad: nivel uno (*Electrical Level One*)
Herrería: nivel uno (*Ironworking Level One*)
Herrería de refuerzo: nivel uno (*Reinforcing Ironwork Level One*)
Instalación de rociadores: nivel uno (*Sprinkler Fitting Level One*)
Instalación de tuberías: nivel uno (*Pipefitting Level One*)
Instrumentación: nivel uno, nivel dos, nivel tres, nivel cuatro (*Instrumentation Levels One through Four*)
Orientación de seguridad (*Safety Orientation*)
Paneles de yeso: nivel uno (*Drywall Level One*)
Seguridad de campo (*Field Safety*)

Acknowledgments

This curriculum was revised as a result of the farsightedness and leadership of the following sponsors:

ABC Pelican Chapter
Bay Ltd.
Bechtel
Bo-Mac Contractors, Ltd.
Cowboyscranes.com
Exelon Generation
Fluor Corp.

KBR, Inc.
Kelley Construction
Mammoet USA
North American Crane Bureau
Orion Marine Group
Southland Safety

This curriculum would not exist were it not for the dedication and unselfish energy of those volunteers who served on the Authoring Team. A sincere thanks is extended to the following:

Ed Burke
Robert Capelli
Anthony Johnson
Richard Laird
Steven Lawrence

Don McDonald
Timothy Prakop
Larry "Cowboy" Proemsey
Joseph Watts
Harold Williamson

A sincere thanks is also extended to the dedication and assistance provided by the following technical advisors:

Monty Chisolm

Keith Denham

Frank Jones

NCCER Partners

American Council for Construction Education
American Fire Sprinkler Association
Associated Builders and Contractors, Inc.
Associated General Contractors of America
Association for Career and Technical Education
Association for Skilled and Technical Sciences
Construction Industry Institute
Construction Users Roundtable
Design Build Institute of America
GSSC – Gulf States Shipbuilders Consortium
ISN
Manufacturing Institute
Mason Contractors Association of America
Merit Contractors Association of Canada
NACE International
National Association of Women in Construction
National Insulation Association
National Technical Honor Society
National Utility Contractors Association
NAWIC Education Foundation
North American Crane Bureau
North American Technician Excellence
Pearson

Prov
SkillsUSA®
Steel Erectors Association of America
U.S. Army Corps of Engineers
University of Florida, M. E. Rinker Sr., School of Construction Management
Women Construction Owners & Executives, USA

Contents

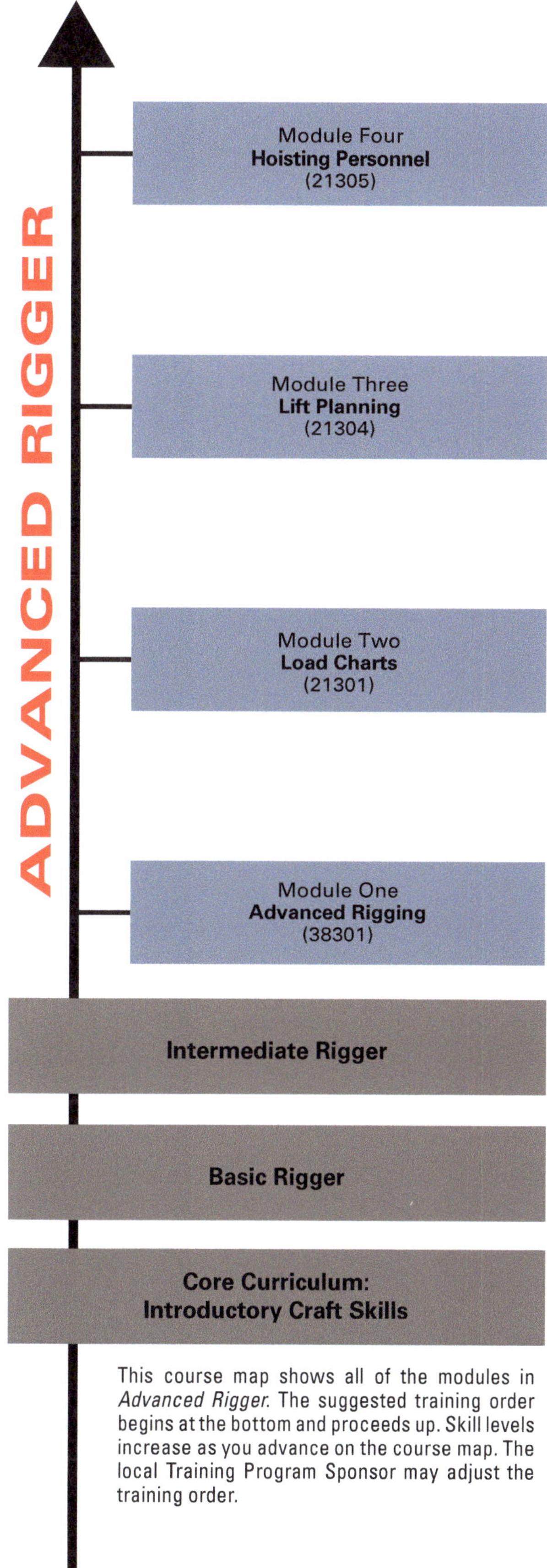

This course map shows all of the modules in *Advanced Rigger*. The suggested training order begins at the bottom and proceeds up. Skill levels increase as you advance on the course map. The local Training Program Sponsor may adjust the training order.

Advanced Rigging

OVERVIEW

Well-planned rigging operations are needed to move many different types of heavy objects vertically and laterally. The safe performance of complex rigging tasks, including those involving cranes, requires a working knowledge of rigging equipment. This module introduces several advanced rigging methods and devices. The complexities of multiple-crane lifts are also examined.

Module 38301

38301
ADVANCED RIGGING

Objectives

When you have completed this module, you will be able to do the following:

1. Explain the concepts of load dynamics and how multi-crane lifts are planned.
   a. Define and calculate load moments.
   b. Define and describe the effects of various factors in crane stability.
   c. Explain the process of determining crane loads for multi-crane lifts.
2. Identify and describe the use of special rigging equipment.
   a. Describe the use and application of cribbing.
   b. Explain how to determine line pull when using inclined planes to move equipment.
   c. Identify and describe the use of spreader bars and equalizer beams.
   d. Explain how to rig and handle reinforcing bar bundles.

Performance Tasks

Under the supervision of your instructor, you should be able to do the following:

1. Select the appropriate spreader bar or equalizer beam for a given load.

Trade Terms

Backward stability
Center of gravity (CG)
Chicago boom
Coefficient of friction (CF)
Combined CG
Cribbing
Dynamics

Forward stability
Impact loading
Inclined planes
Momentum
Non-centered lift
Pendulum effect
Rebar

Industry Recognized Credentials

If you are training through an NCCER-accredited sponsor, you may be eligible for credentials from NCCER's Registry. The ID number for this module is 38301. Note that this module may have been used in other NCCER curricula and may apply to other level completions. Contact NCCER's Registry at 888.622.3720 or go to **www.nccer.org** for more information.

Contents

1.0.0 LOAD DYNAMICS

Objective

Define and explain the concepts of load dynamics.

    a. Define and calculate load moments.
    b. Define and describe the effects of various factors in crane stability.
    c. Explain the process of determining crane loads for multi-crane lifts.

Trade Terms

Backward stability: In crane operations, the measure of crane steadiness in relation to the tipping axis on the opposite side of the crane from the boom.

Center of gravity (CG): The point where one can assume all of an object's mass, and therefore its weight, is concentrated. The concept is useful for determining stability, a load's balance point, and leverage.

Combined CG: The sum of the crane's and load's CGs at a position along an imaginary line connecting the two CGs.

Dynamics: The response of objects to the application of forces, moments, or torques.

Forward stability: In crane operations, the measure of the crane's steadiness in relation to the tipping axis on the same side as the boom.

Impact loading: Sudden forces acting on an object due to collisions or other dynamic events.

Momentum: A property of a moving object that is directly proportional to both its mass and speed.

Non-centered lift: Any lift attempted when the hoist line is not vertical or the hook is not directly above the load's center of gravity.

Pendulum effect: The dynamic effects of a swinging object on the support from which it is suspended.

Cranes used on the jobsite are valuable construction tools. They operate according to the basic principles of leverage and stability. Leverage, or moment, is the effect of a force applied to a lever to move a load. Crane manufacturers design cranes so they can exert the necessary leverage to move loads while remaining stable. Stability describes the ability of an object to resist overturning or rotating due to gravity.

1.1.0 Rotational Forces and Moments

Other modules in this curriculum cover the basic concepts of leverage and moments. This section provides a review of those concepts and explains how they apply to rigging practices.

Moments and torques are types of rotational forces. Depending on the application, a mechanical moment is the effect of a force acting on a lever arm at a certain distance from a pivot. Torque is often thought of as the force exerted by a turning shaft at a certain distance on structures surrounding it. In other applications, both *moment* and *torque* can mean the same thing. Moments significantly contribute to the overall dynamics of crane operations.

1.1.1 Definition of Moment

A more exact definition of a moment is the product of a force acting perpendicularly to an arm of some length attached to a point around which it rotates (*Figure 1*). The formula for mechanical moment is:

$$M = D \times F_{Perp}$$

Where:

M = symbol for a mechanical moment
D = distance or length of the moment arm between the position where the force acts and the pivot point
F_{Perp} = force applied perpendicular to the moment arm

> **NOTE**
>
> Mechanical forces are pushes or pulls. Weight is the force exerted by gravity as it pulls downward on an object. The weight of an object pushes or pulls down on whatever is supporting it.

1.1.2 Equal Moments and Balance

Riggers need to understand mechanical moments as they apply to cranes. There are typically two moment arms to consider—the crane's and the load's. Crane operators usually imagine these moment arms being rigidly connected at the pivot point or fulcrum, similar to a teeter-totter. *Figure 2* shows equal weights (W) placed on a lever at equal distances (X) from the fulcrum. If the weights are both 500 pounds and their moment arms are both 20 feet, then their moments (M) are:

$$M_1 = X_1 \times W_1$$
$$M_1 = 20 \text{ ft} \times 100 \text{ lb}$$
$$M_1 = 2{,}000 \text{ ft-lb}$$

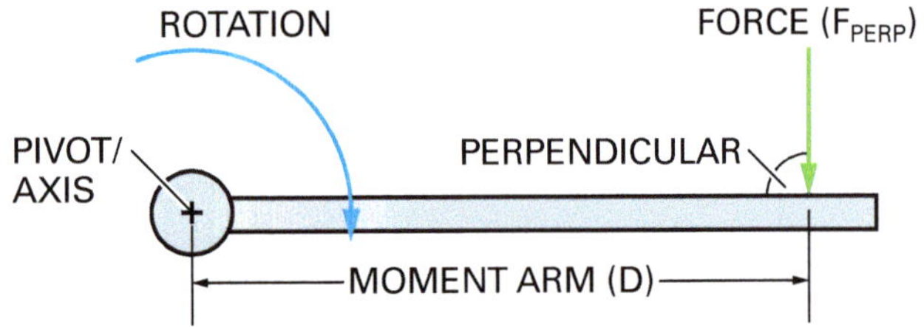

Figure 1 Description of a mechanical moment.

> **NOTE**
>
> When performing multiplication and division, units combine as in this example. Multiplying feet times pounds results in the unit *foot x pound*, or *foot-pound*. The unit for mechanical moment and torques is the *foot-pound*, abbreviated *ft-lb*. In metric units, it is the *kilogram-meter (kg-m)*.

Similarly:

$$M_2 = X_2 \times W_2$$
$$M_2 = 20 \text{ ft} \times 500 \text{ lb}$$
$$M_2 = 10,000 \text{ ft-lb}$$

> **NOTE**
>
> The calculated moments of the weights in *Figure 2* are correct as long as the bar is horizontal and the full weights act perpendicular to the bar.

Moment M_1 creates a counterclockwise (CCW) rotation while M_2 causes a clockwise (CW) rotation. Because the two moments are acting on the same lever simultaneously, they cancel each other out and there is no motion. The lever is balanced.

In *Figure 3*, the distances are again equal, but the weights are unequal. If the fulcrum remains at the same position under the lever, the heavier side will cause the lighter side to rise. While this seems to be normal behavior, why does it happen? Assume again that both moment arms are 20 feet. Weight 1 is 1,000 pounds and Weight 2 is 500 pounds. Calculate the two moments:

$$M_1 = X_1 \times W_1$$
$$M_1 = 20 \text{ ft} \times 1,000 \text{ lb}$$
$$M_1 = 20,000 \text{ ft-lb (CCW)}$$

$$M_2 = X_2 \times W_2$$
$$M_2 = 20 \text{ ft} \times 500 \text{ lb}$$
$$M_2 = 10,000 \text{ ft-lb (CW)}$$

The CCW moment is twice the size of the CW moment. As shown in *Figure 3*, the lever begins to rotate in the CCW direction. How far will it rotate? Rotation will continue until W_1 can no longer fall under the influence of gravity.

In the case of a teeter-totter, it is the ground that usually stops the motion if a rider's legs do not. The ground's surface is rigid; it pushes up on the heavy end with exactly the same amount of force as the extra weight pulling down. Since there is no longer any difference in moments, and the lever can't move any farther, rotation stops.

You have seen how adding weight to one side can unbalance a lever. How can you balance a lever given two different weights? In other words, how can you produce two equal moments from unequal weights? Assuming a physical lever with the fulcrum at its middle, and assuming the lighter weight is at the end of its lever arm, the only way to equalize the mechanical moments is to shorten the moment arm of the heavier weight—its distance from the fulcrum.

The goal of this example is to make the moments of both weights equal. In equation form:

$$X_1 \times W_1 = X_2 \times W_2$$

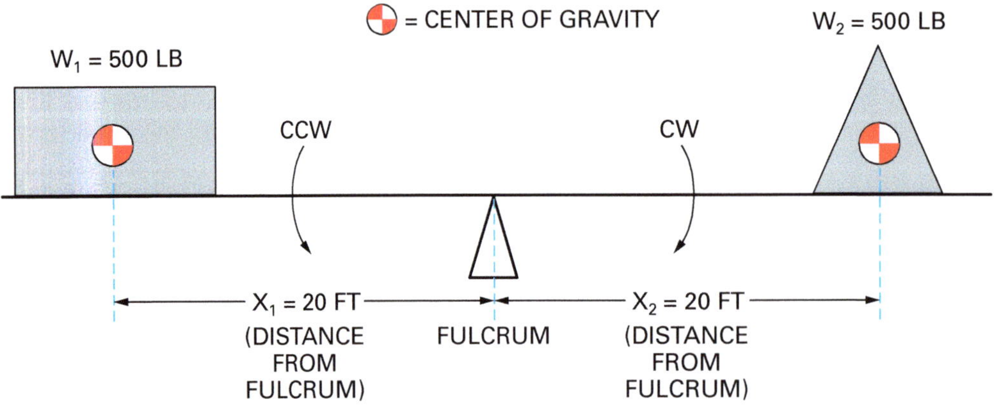

Figure 2 Loads in balance.

What you know:

$$W_1 = 1{,}000 \text{ lb}$$
$$W_2 = 500 \text{ lb}$$
$$X_2 = 20 \text{ ft}$$
$$X_1 = ?$$

Solve for X_1 by dividing both sides by W_1 to give the following format:

$$X_1 = \frac{X_2 \times W_2}{W_1}$$

$$X_1 = \frac{20 \text{ ft} \times 500 \text{ lb}}{1{,}000 \text{ lb}}$$

$$X_1 = \frac{10{,}000 \text{ ft-lb}}{1{,}000 \text{ lb}}$$

$$X_1 = 10 \text{ ft}$$

Since W_1 is twice the weight of W_2, it is reasonable that its moment arm is half that of the other weight. The now-equal moments act in opposite directions around the fulcrum, so the lever is balanced. *Figure 4* shows the result of moving W_1 towards the fulcrum to the 10-foot mark, again resulting in balance.

1.1.3 Moments and Cranes

Imagine a crane and its load acting like a teeter-totter. This is relatively easy if you simplify its shape, as shown in *Figure 5*. The side to the left of the bend in the lever represents the crane base, upperworks, and counterweight. The right side of the lever angles up, simulating a crane's boom. A weight representing the suspended rigging and load hangs from the boom tip. Crane operators refer to the fulcrum of the crane-lever model as the *tipping axis*, as identified in the figure.

A mechanical moment is the product of its moment arm and the part of the force perpendicular to the moment arm. In *Figure 5*, the crane's moment arm is the horizontal distance between its center of gravity (CG) and the tipping axis. While this isn't completely accurate, you will see that this definition works.

The load's moment arm is a little more complicated to explain. Physically, the load moment arm is the distance between the tipping axis on the ground and the boom tip, as shown in the figure. The boom tip is the point where the suspended weight acts on the crane. But the load is pulling down at an angle to the boom. The force truly producing the load's moment is only a portion of

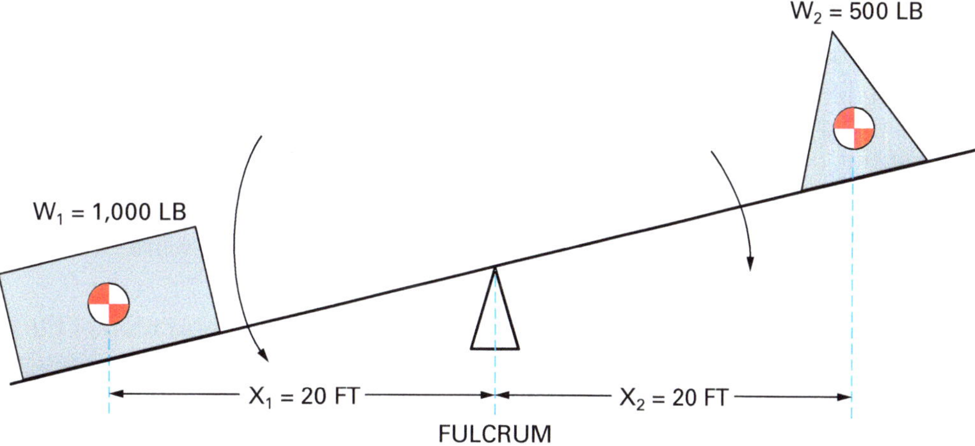

Figure 3 Unbalanced loads.

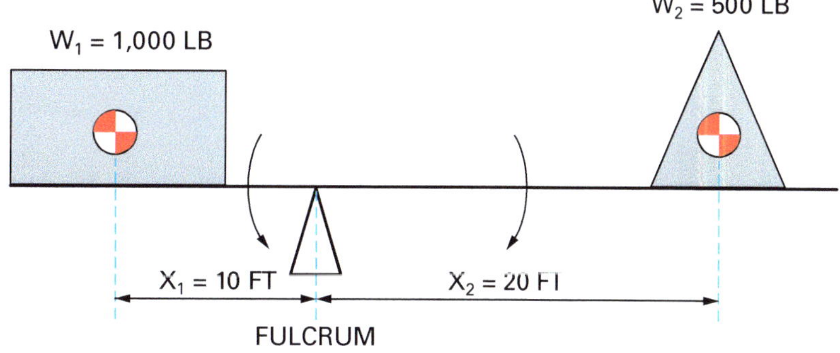

Figure 4 Balancing two loads of unequal weight.

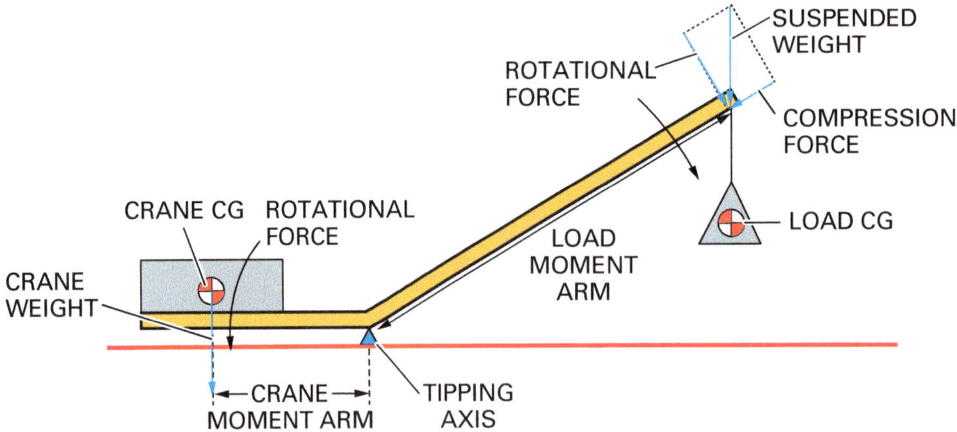

Figure 5 A crane and its load viewed as a lever.

the suspended load weight—that part acting perpendicularly to the boom. This is the rotational force on the boom. The other part of the load's weight works to compress the boom along its length.

Though these forces on the boom are real, treating a load moment like this greatly complicates the operators' and riggers' understanding of its effect on the crane. It's simpler and equally correct to measure the load moment arm as the shortest horizontal distance from the tipping axis on the ground to a point directly under the CG of the suspended load (*Figure 6*). This way, you can use the entire suspended load as the rotational force, which acts perpendicularly to the load's moment arm parallel to the ground. This approach also helps to explain the effect of the load operating radius on crane stability.

A mobile crane is stable when the crane's moment is greater than the load's moment. A crane's stability decreases as the load radius increases or it lifts heavier loads—anything that increases the load's moment or decreases the crane's moment. If the load's moment exceeds the crane's moment, the crane will rotate around its tipping axis—it tips. If both moments are equal, the crane's stability is in jeopardy. Theoretically, in this situation, a worker could tip the crane by simply pushing up at a point near the rear counterweights of the crane.

The amount of rotational force on the crane side of the tipping axis must always be greater than the rotational force on the side of the load. This ensures that the crane will not tip during lifting operations. When cranes are designed, the gross load that they are capable of lifting at a given load radius is specified on a load chart. The chart takes into consideration the configuration of the crane and the centers of gravity that move as it performs a lift.

1.2.0 Crane Stability

Crane stability depends on many factors. For example, the CG of the crane moves as its configuration changes, resulting in a change to the crane's stability under load. The following aspects of crane stability are discussed in the sections that follow:

- The configuration of the crane and its resulting CG
- The quadrant of operation of the crane

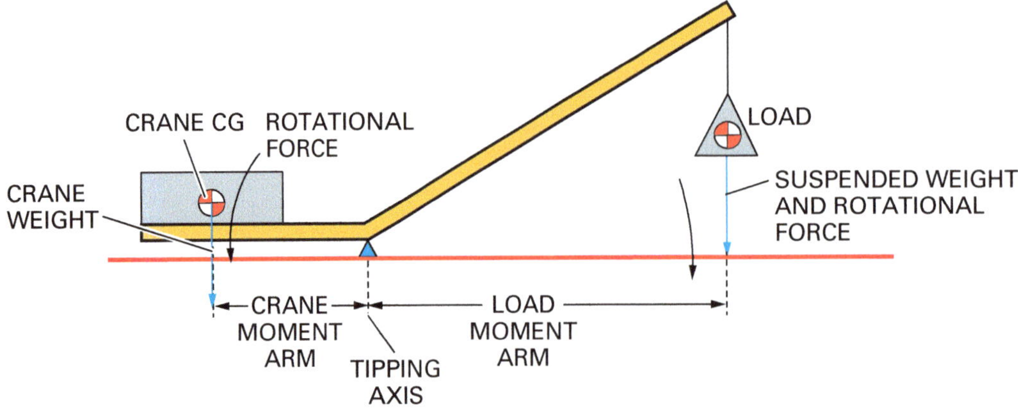

Figure 6 Mechanical moments affecting crane stability.

NCCER – *Advanced Rigger*

- Forward stability
- Backward stability
- Non-centered lifts
- Environmental factors

1.2.1 Crane Configuration

A mobile crane is manufactured with hundreds of subcomponents. Each component has a weight and a location within the crane's structure. Many components have predictable weights and positions. For example, the crane's engine, upperworks enclosure, operator's cab, carrier, wheels, and crawlers are relatively constant weights. The locations of some of these can change as the upperworks rotates, while others are fixed. Counterweights and many other crane components, however, can vary in either weight, position, or both.

Each part of a crane has its own CG—a point in space where the weight of the part is concentrated (*Figure 7*). If one averages the component CGs at any given instant in time, the result is the crane's overall CG, indicated by the larger CG symbol in the figure. Adding counterweight to the rear of the upperworks moves the crane's CG toward the counterweight (rear of the crane). Lowering the boom lowers the boom's CG and moves it away from the crane. This has the effect of moving the crane's overall CG toward the boom, and it also lowers it slightly. Notice that a change in the boom's configuration—adding or removing lattice sections, extending or retracting a telescopic boom, and installing/removing a jib extension—occurs on the opposite side of the tipping axis from the crane's CG. Therefore, altering the configuration of the boom and load-handling components can greatly affect the crane's stability.

Figure 7 is an engineering drawing that shows how to locate centers of gravity using common horizontal and vertical reference points.

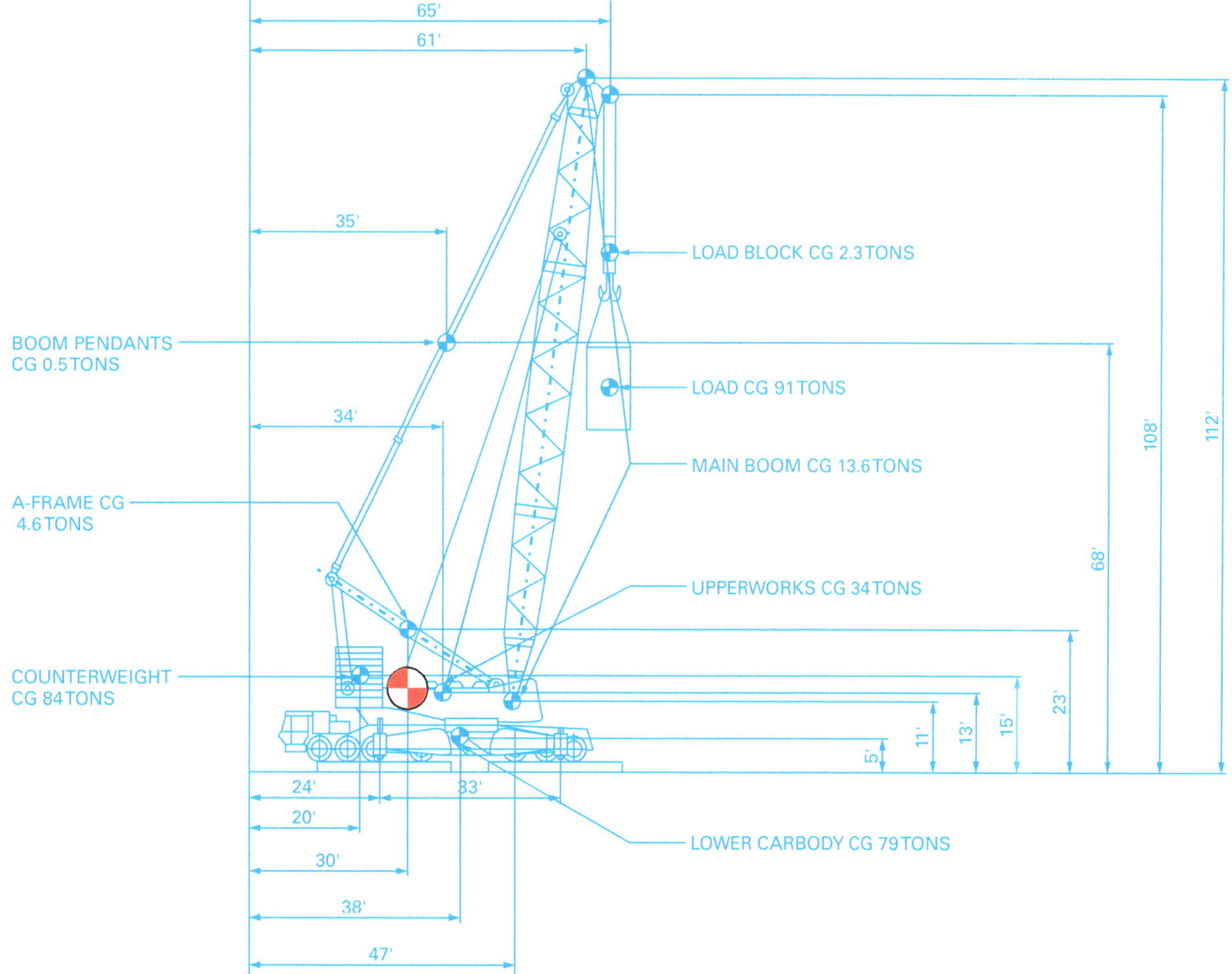

Figure 7 Contributions of component CGs to the crane's overall CG.

1.2.2 Quadrants of Operation

As a crane's upperworks swings around, the distance of the crane's overall CG from the applicable tipping axis changes dramatically. This is because manufacturers typically build cranes to be longer than they are wide. This shape allows them to fit on a roadway or trailer for transport from job to job. Whether on crawlers, tires, or outriggers, the rectangular-shaped footprint tends to result in more stability of the crane in the front-rear directions than from side-to-side.

Manufacturers define the operating quadrants for each of the cranes that they produce. *Figure 8* shows typical examples for the various carrier types, with and without outriggers. Notice that the points of contact with the supporting surface define the operating quadrants as well as the tipping axis.

The crane manufacturer identifies the authorized quadrants of operation in the operator's manual and load charts, and specifies any limitations while operating in those areas. The operating quadrants are:

- *Over-the-front* – An area located between the front outriggers or crawlers when facing in the forward direction of travel.
- *Over-the-rear* – The area between the rear outriggers or crawlers when facing the rearward direction of travel. A truck-bodied crane on outriggers typically has its best stability in this quadrant.
- *Over-the-side* – A position located between the front and back outriggers or over the crawler on either side.
- *360-degree* – The circular swing area that would encompass all the other quadrants.

1.2.3 Centers of Gravity and Forward Stability

The motions of the various components and their CGs during crane operation dynamically change the crane's overall CG. The shortest horizontal distance of the crane's CG from the tipping axis in the direction of the load determines the crane's moment arm. At the same time, swinging the crane with a load or extending/retracting the boom continually changes the load's moment arm relative to the tipping axis. To ensure stability, the operator must control the crane so that its moment is always larger than the load's moment. It is sometimes easier to understand a crane's forward stability by visualizing the location of the crane and load's combined CG compared to the tipping axis. The combined CG is the sum of the crane's and load's CGs at a position on an imaginary line connecting the two CGs (*Figure 9*). As long as the combined CG remains on the crane side of the tipping axis, the crane remains stable. Any configuration change or operation that moves the combined CG toward the tipping axis decreases crane stability. Such changes may include:

- Reducing counterweight
- Setting up a boom extension
- Booming down; telescoping out
- Rotating the upperworks
- Changing the luffing-jib offset
- Suspending a load

If the combined CG crosses the tipping axis, the crane will tip toward the load, usually with catastrophic results.

The following example illustrates the usefulness of the combined CG concept.

Example One:

Refer to *Figure 10*, which depicts a crane and load during a lift. Assume it is operating in the rear quadrant. The diagram provides their weights and moment arms relative to the rear tipping axis. For simplicity, assume that the boom is perpendicular to the tipping axis.

The total weight of the crane and load (W_T) is simply the sum of the crane's weight (W_C) and the weight of the load (W_L):

$$W_T = W_T + W_L$$
$$W_T = 4,000 \text{ lb} + 500 \text{ lb}$$
$$W_T = 4,500 \text{ lb}$$

This is the weight of the combined CG. The location of the combined CG requires using the mathematical weighted-average method, described in the following steps:

Step 1   Assign the crane's moment arm (X_C) negative distance values as measured from the tipping axis, and assign the load's moment arm (X_L) positive values.

Step 2   Calculate the crane's and load's moments.

Crane's moment:

$$M_C = X_C \times W_C$$
$$M_C = -8 \text{ ft} \times 4,000 \text{ lb}$$
$$M_C = -32,000 \text{ ft-lb}$$

Load's moment:

$$M_L = X_L \times W_L$$
$$M_L = 20 \text{ ft} \times 500 \text{ lb}$$
$$M_L = +10,000 \text{ ft-lb}$$

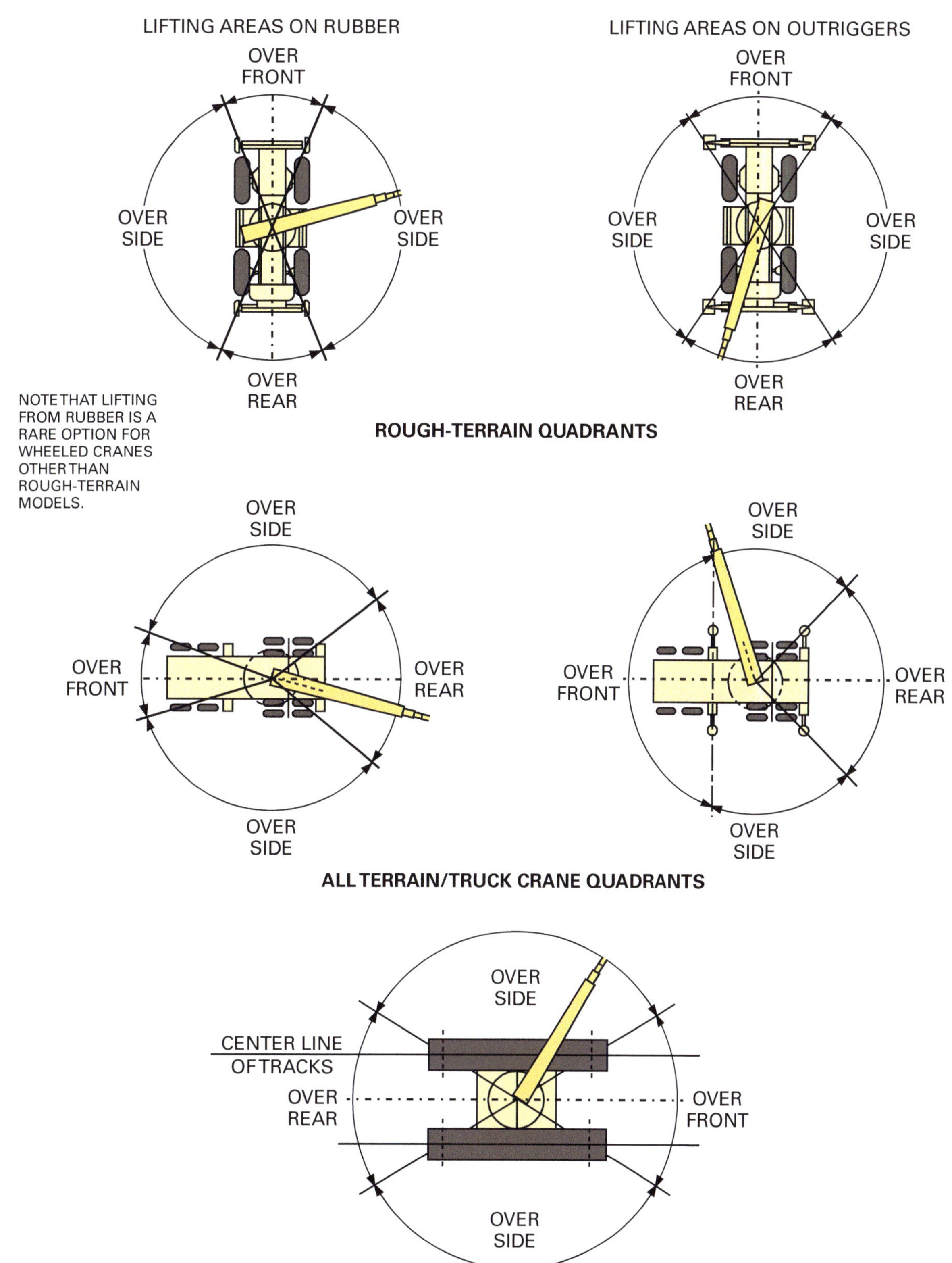

Figure 8 Mobile crane operational quadrants.

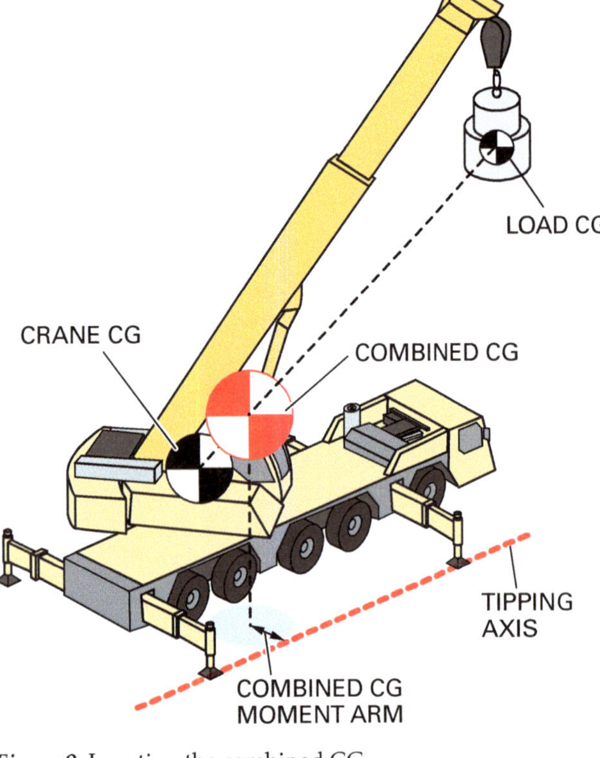

Figure 9 Locating the combined CG.

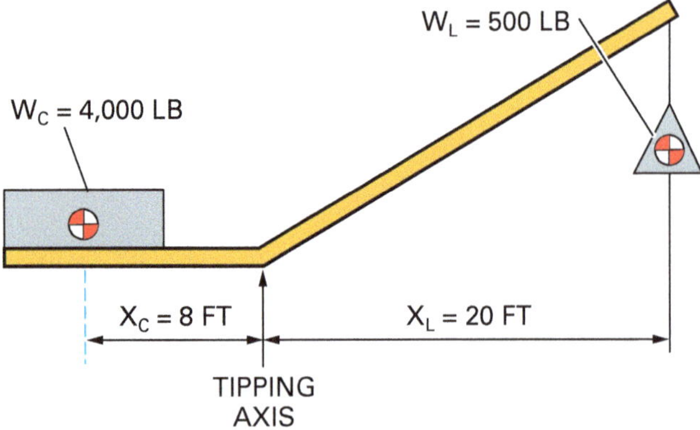

Figure 10 Stability Example One.

Step 3 Add the two moments together to find the combined moment.

$$\text{Combined moment} = M_C + M_L$$
$$(-32,000 \text{ ft-lb}) + (+10,000 \text{ ft-lb})$$
$$-22,000 \text{ ft-lb}$$

Step 4 Divide the combined moment by the combined weight to calculate the combined moment arm (X_{Comb}).

$$X_{Comb} = -22,000 \text{ ft-lb} \div 4,500 \text{ lb}$$
$$X_{Comb} = -4.9 \text{ ft}$$

The combined CG's moment arm is negative, indicating that the combined CG is on the crane side of the tipping axis. The crane is stable in this example.

CAUTION

The example calculations illustrating the location of the combined CG are provided merely to demonstrate the concept. Riggers and crane operators are not normally required to perform calculations such as this in the field. Operators should rely only on the appropriate load charts and other manufacturer-provided information to determine crane stability.

1.2.4 Backward Stability

The crane operator must be concerned with reverse, or backward stability, as well. The dynamics are only slightly different from those of forward stability. The concern is to keep the combined CG of the crane, including its load, if applicable, from moving past the tipping axis on the opposite side of the crane from the load/boom. If this situation occurs, the crane will tip over backwards.

Several factors can contribute to backward tipping, such as when the boom applies a sudden upward moment to the crane. This can occur by either suddenly stopping the boom while raising it or by suddenly releasing the load. In either case, the moving boom transfers its momentum to the crane through its actuators and hinge points. Excessive boom recoil can drive a boom against its mechanical stops. This creates a lever action that can tip the crane backward, as shown in *Figure 11*.

The same lever action occurs when the operator accidentally raises the boom too far into the boom stops or actuator limits. The upward jerk in the boom, especially at high boom angles, can momentarily shift the crane's combined CG past the point of no return, resulting in the crane tipping backward.

Another factor that can affect backward stability is the orientation of the crane and boom to unlevel ground. If the operator raises a long boom to a high angle, this action decreases the distance between the crane's CG and rear tipping axis, reducing the crane's stability. If the crane then travels up an incline in this configuration, the crane's CG can move past the rear tipping axis, leading to a crane tipping accident. Operating a crane on unlevel ground requires caution and planning beforehand to anticipate the potential impact of these conditions.

Another CG Example

Assume that the crane operator swings the boom and load to a different quadrant, such as over-the-side. The crane and load weights remain the same as in the previous example, as shown here. However, because the side-quadrant tipping axis now applies, the moment arms have changed.

Work through the same calculations performed for Example One. First, calculate the crane's and load's moments:

- *Crane's moment:*

$$M_C = X_C \times W_C$$
$$M_C = -3 \text{ ft} \times 4{,}000 \text{ lb}$$
$$M_C = -12{,}000 \text{ ft-lb}$$

- *Load's moment:*

$$M_L = X_L \times W_L$$
$$M_L = 24 \text{ ft} \times 500 \text{ lb}$$
$$M_L = +12{,}000 \text{ ft-lb}$$

Then, calculate the combined moment:

$$\text{Combined moment} =$$
$$M_C + M_L$$
$$(-12{,}000 \text{ ft-lb}) + (+12{,}000 \text{ ft-lb})$$
$$0 \text{ ft-lb}$$

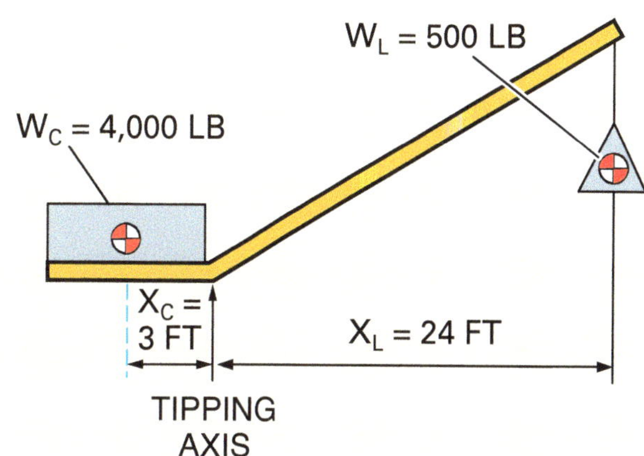

Lastly, calculate the combined moment arm (X_{Comb}):

$$X_{Comb} = 0 \text{ ft-lb} \div 4{,}500 \text{ lb}$$
$$X_{Comb} = 0 \text{ ft}$$

The combined CG is directly over the tipping axis. The crane operator should not allow this situation to occur, as the crane is very unstable. Any small deviation in the load or crane moment could move the combined CG beyond the tipping axis, causing the crane to tip toward the load. Moving a load from one operating quadrant to another can have a dramatic effect on crane stability.

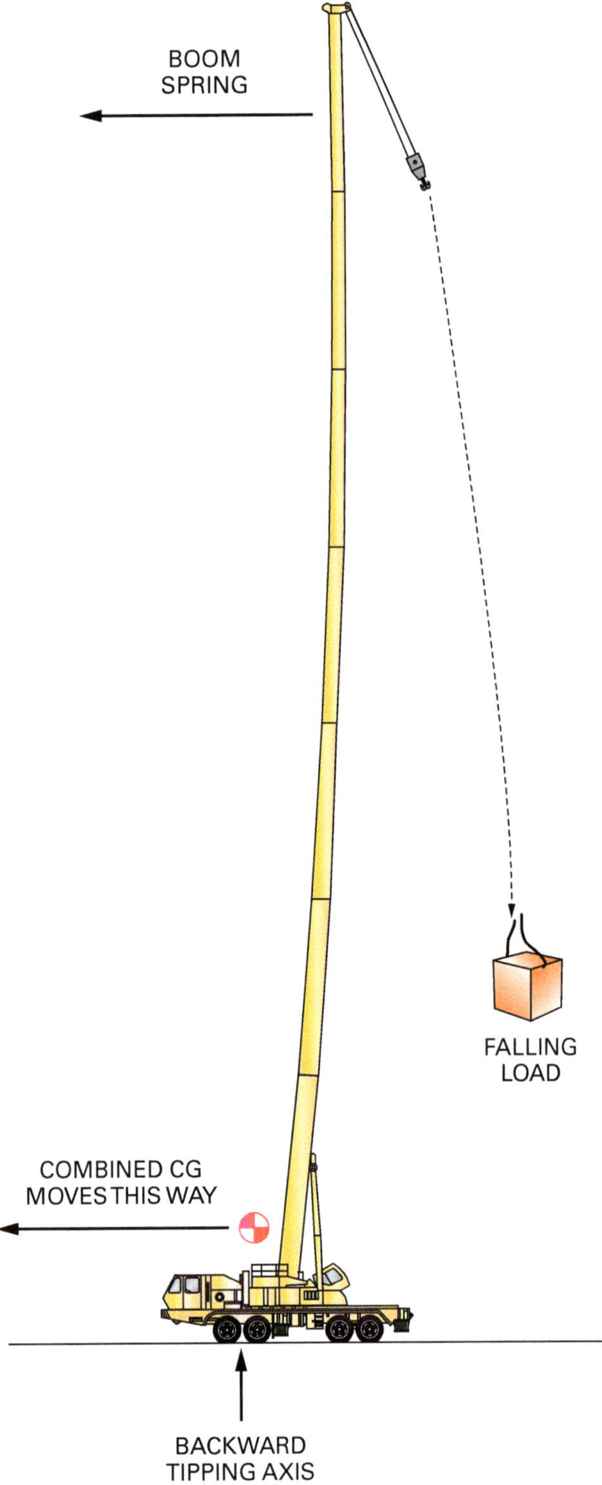

Figure 11 Backward tipping due to boom spring.

Rough-terrain (RT) cranes are prone to backward tipping in certain situations. Crane manufacturers are making RT cranes larger and with heavier counterweights. This can create a problem if the operator doesn't follow the manufacturer's recommendations when swinging the crane boom from over-the-front to over-the-side on rubber. Failure to compensate for heavy counterweights and short crane moment arms (booms) when swinging from one quadrant to another could cause the crane to go over backwards. As mentioned earlier, manufacturers typically build cranes to be longer than they are wide, which results in more stability in the long dimension. A narrow carrier and heavy counterweights can combine to cause backward tipping, especially at high boom angles with no load.

1.2.5 Non-Centered Lifts and the Effects on Load Radius

A non-centered lift can affect a crane's stability and structural integrity. If the crane does not lift the load with the load line vertical and the hook centered directly over the load's CG, several problems can result. Side-loading of the boom can stress the boom and its suspension system, leading to structural failure. If the load is farther from the crane base than the boom tip, the hoist line will carry a higher tension than the actual weight of the load. The effect is similar to how the tension in an angled sling under load is greater than for a vertical hitch. The hoist line tension could exceed the permitted load at the existing boom angle and load chart operating radius.

Attempting to lift a load not directly under the boom tip will cause the load to drag until it is suspended. The resulting friction adds to the hoist line tension, and the load will likely be damaged as well.

Another problem that occurs when suspending a non-centered load is the pendulum effect. As the load clears the ground, its CG will swing through the vertical line below the boom tip to the opposite side, like a pendulum. When the load is at the extremes of the swing, the tension in the hoist line is greater than the weight of the load alone. The swinging action creates a very unstable condition, resulting in wide variations in boom side-loading, hoist line tension, and/or the measured load radius. The pendulum effect is actually put to good use in ball demolition work, but it must be performed with great care. A swinging ball, or any other swinging load, can strike the boom and cause catastrophic damage.

Swinging the boom too rapidly during a lift can cause similar problems, with the load following a path having a larger radius than the boom tip (*Figure 12*). This action can add significant load to the boom unaccounted for in the load chart. Attempting to stop such a motion without considering the load's momentum can severely side-load the boom.

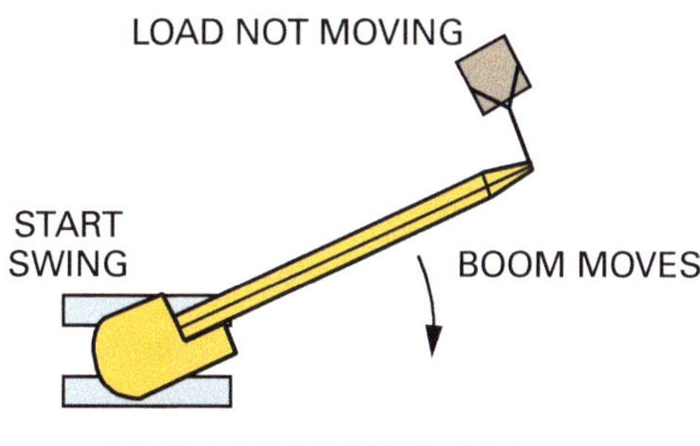

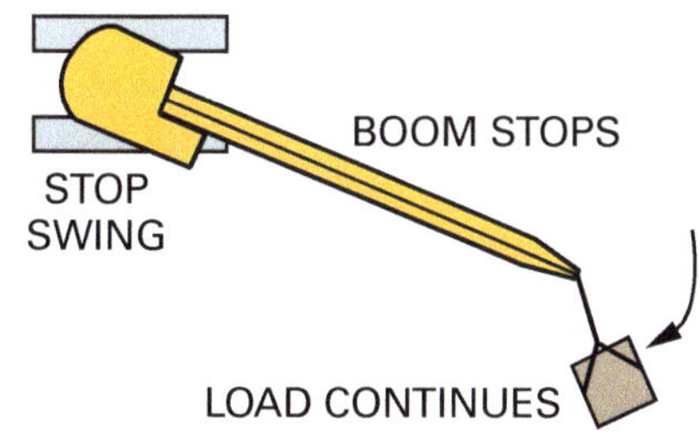

Figure 12 Dynamic loading during swings.

These types of motions can overload the crane, causing either a structural failure or a tipping incident. Signal persons must be able to signal to the operator how to stop a swinging load, using the following general guidelines:

- For a side-to-side swing, move the boom in the direction of the swing as it reaches the apex of the swing.
- For front-to-back swings, boom up on the near side and boom down on the far side as the load approaches the apex of the swing.
- If the swing is circular, damp the swing front-to-back or side-to-side first, then damp the other swing.

1.2.6 Wind

One of the most overlooked stability factors affecting cranes is the wind (*Figure 13*). High wind speeds can dramatically affect a crane and its load. Almost all load charts require reducing the load chart ratings under windy conditions, and they may also recommend a shutdown wind speed.

In many cases, when the wind speed exceeds 30 mph (48 kph), it is advisable to stop operations.

Wind affects both the crane and the load, changing the crane's load moment and thus the rated capacity of the crane. Operators need to use a great deal of care, even when lifting with moderate wind speeds that a crane's load chart allows, especially if the winds are gusty. The crane operator must be aware of the precise out-of-service wind speed for the crane as specified by the manufacturer. If this information is ignored, equipment damage, injuries, and/or loss of life can occur. Operators must consult the wind speed charts as necessary and follow them without fail.

Table 1 is an example of a wind speed chart for a given crane. Note that some cities and counties have specific regulations regarding wind speed that are stricter than the manufacturer's requirements. If a jobsite or locality imposes a wind speed limit that differs from the manufacturer's charts, the lower value is the limit to use.

> **CAUTION**
>
> The wind chart shown in *Table 1* is for a specific crane model. The information provided is not applicable to all cranes. Always check and follow the manufacturer's wind chart speed for the specific crane in use.

Crane operators and riggers must consider the wind-catching surface of the load whenever wind is present. Control of the load can easily be lost. A 20 mph (32 kph) wind exerts a pressure of only 1.2 lb/ft² (5.9 kg/m²) on a flat-surfaced load, for example. The force exerted on a common sheet of plywood is 38 lb (17 kg). Under these conditions, only loads having a large sail area (surface area that would catch the wind) may require the crane's capacity to be reduced in accordance with the manufacturer's guidance. At 30 mph (48 kph), however, the wind exerts a pressure of 2.6 lb/ft² (12.7 kg/m²) on the same flat surface area, more than double that of a 20-mph wind. This results in an 84 lb (38 kg) force exerted on a common sheet of plywood. This much wind is enough to cause load control problems.

Forward stability is the critical consideration when the wind is coming from behind the boom. It applies a force to both the boom and load that adds to the tipping moment of the crane. This has the same effect as adding load to the hook.

Backward stability is the critical factor when the wind is from the front, particularly when the boom is at or approaching the maximum boom angle. This has the same effect as reducing the load on the hook and exerting a backward

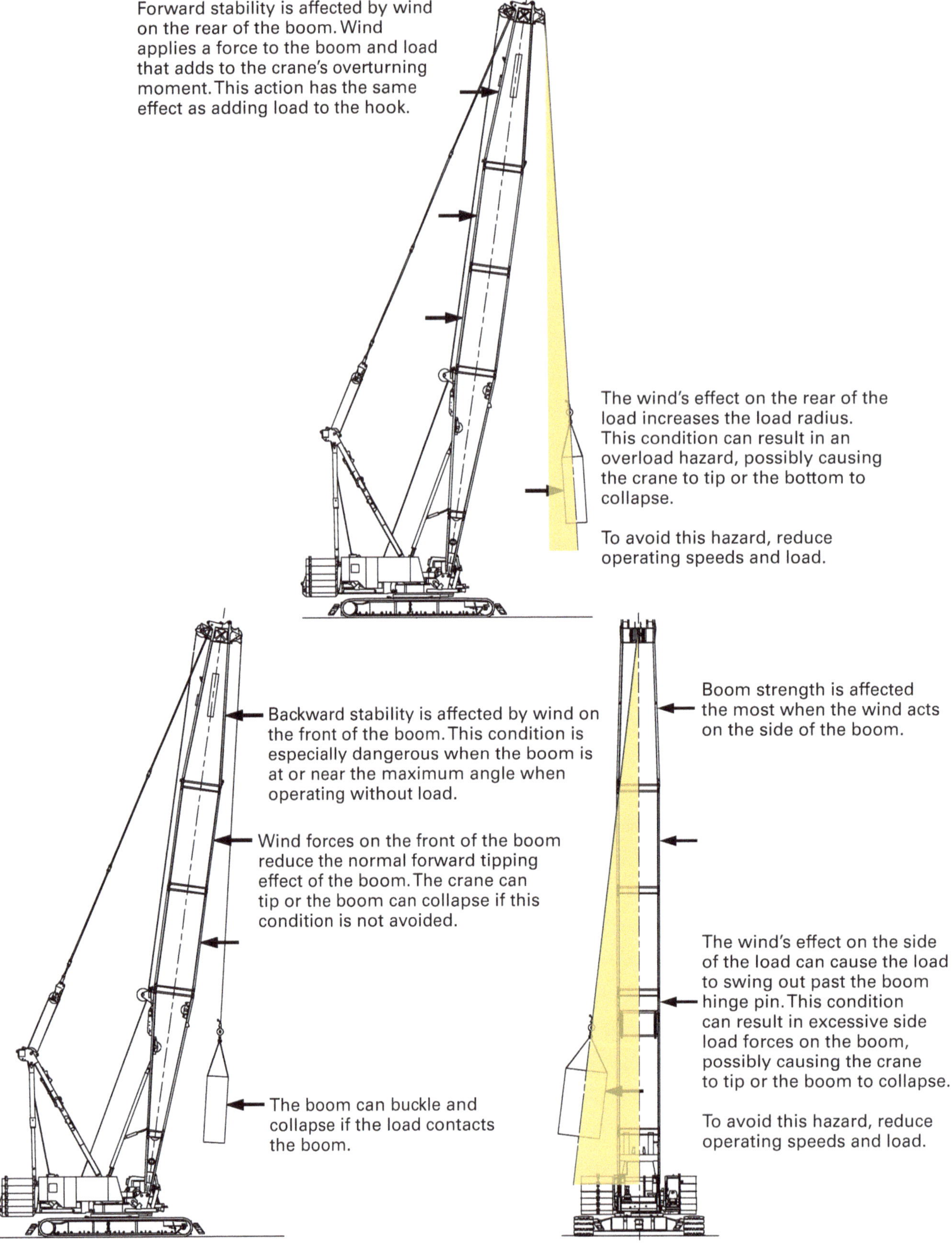

Figure 13 Wind and its effect on crane stability.

Table 1 Example of a Wind Speed Chart

Boom and Boom + Jib Lengths up to 250'	
Description	**Allowable Windspeeds in Miles Per Hour (mph)**
1. Normal Lifting Operation. (See Capacity Charts.)	0–20 mph
2. Reduced Operation. Capacities must be reduced by 20%.	21–30 mph
3. Reduced Operation. Capacities must be reduced by 40%.	31–40 mph
4. Reduced Operation. Capacities must be reduced by 70%.	41–45 mph
5. No Operation. Store attachment on ground.	Over 45 mph
Boom and Boom + Jib Lengths Greater than to 250'	
1. Normal Lifting Operation. (See Capacity Charts.)	0–20 mph
2. Reduced Operation. Capacities must be reduced by 35%.	21–30 mph
3. Reduced Operation. Capacities must be reduced by 60%.	31–40 mph
4. Reduced Operation. Capacities must be reduced by 70%.	41–45 mph
5. No Operation. Store attachment on ground.	Over 45 mph

moment using the boom as a lever. The wind forces on the boom reduce the forward moment normally provided by the boom by lifting it higher and closer to the tipping axis.

1.2.7 Impact Forces

Another equally dangerous dynamic effect is impact loading, also known as *dynamic loading*. An impact load is the force exerted on a crane by a moving object when rapidly stopping or slowing a load's descent. A large, short-duration force is required to eliminate the momentum of the moving object while stopping it. Impact loading occurs, for example, when the hoist drum brake suddenly snubs a free-falling load. Snatching a load off the ground will also cause an impact load to occur.

Impact loading causes an instantaneous increase in the load supported by the crane. When the operator releases a load and allows it to drop freely, the boom will initially recoil upward as the drop begins. However, when the operator brakes or locks the hoist drum to stop the descent, the crane suddenly bears the full weight of the load, plus the impact loading to stop its movement. The result can be the overturning of the crane or structural failure of the boom.

Similar changes to effective load weights (and the load moments) that result from impact loading occur during a large acceleration or deceleration of a load. A rapid hoist acceleration will produce a hook load larger than the actual load weight. If there is a sudden release of a lifted load, the action can cause the boom to recoil, rapidly releasing the boom-flex that normally occurs under load. If this happens at a high boom angle, the crane can topple over backwards, as described previously. The crane operator must remember to gradually transfer the load's weight to the crane when hoisting. When lowering, the operator should gently place the load to allow boom deflection and pendant stretch to gradually return to normal.

1.3.0 Multiple-Crane Lifts

Multiple-crane lifts are always critical lifts, requiring detailed written plans and, when required, written procedures. Planners use multiple cranes when the physical dimensions, characteristics, mass, or placement of the load requires the use of more than one crane. Such lifts need detailed planning, because the weight held by any one of the cranes can change as the lift progresses. As a general rule, lift planners and operators must avoid allowing any one crane to exceed 75 percent of its load chart capacity.

Planners should consider the following factors when planning multiple-crane lifts:

- Qualified, experienced planners must plan a multiple-crane lift and a qualified lift director must implement the plan.
- If one of the cranes does exceed 75 percent of its net capacity, then the operation requires extra engineering evaluations.
- Planners must determine exactly how much of the load's weight each crane will carry at every point during the lift. The rigging design must divide the load as planned.
- During any movement of the load, the line, swing, and boom speeds of the cranes must work together to produce a smooth, even motion.
- Swing and boom motions should be minimized.
- Whenever possible, the cranes should not travel with the load.
- Participants should conduct a dry run without a load or with a test load, if possible, before lifting the actual load.
- Communications by radio are preferred when hand signals are considered less effective.
- Only one signal person should direct and control the operation unless that is not feasible. In that case, participants must develop and test specific communication procedures involving multiple signal persons.
- Operators must keep hoist lines vertical at all times. When the lines are not vertical, the cranes are transferring load to each other; one crane may side-load the other or exceed its planned capacity.

Figure 14 shows two 750-ton crawler cranes performing a coordinated critical lift to position prefabricated steel structures. This lift required extensive site preparations and a detailed engineering review.

Whenever two or more cranes are needed to lift a load, lift planners must carefully choose the location of each crane's load block so that the cranes safely share the load. Because cranes change their boom position and load radius as the lift progresses, net capacities are constantly changing. The actual load a crane carries may also be more or less than the other crane(s) because the load is not symmetrical or its CG is off-center. The following scenario is a practical example of a multiple-crane lift plan.

Planners must determine the weight lifted by two cranes moving a large 180,000-pound structural member (*Figure 15*). Since the load's CG is off-center, the structural member would impose different weights on the two cranes if lifted from its ends. For critical lifts, load planners must document the maximum percentage of net capacity for each crane that will occur during the lift.

Steps 1 through 9 outline the approach for determining the attachment points for lifting an object with two cranes. For this example, assume Crane One's minimum load chart capacity during the lift is 156,000 lb and Crane Two's is 90,500 lb. The load's weight is 180,000 lb. The following procedure determines the hook-block positions on the load and the percent crane capacities.

Step 1   Determine, from Crane One's load chart, the minimum net capacity that it will have during the whole operation.

Minimum net capacity of Crane One:
156,000 lb − 500 lb (beam clamp) = 155,500 lb

> **NOTE**
>
> When lifting from the main boom or an extension, consider all factors: configuration, quadrant of operation, boom length, boom angle, load radius, weight of rigging, and all capacity deductions. Module 21301 of this curriculum presents detailed instructions for the use and interpretation of mobile crane load charts.

> **NOTE**
>
> When using two cranes, measure the load radius of each crane to its attachment point on the load, not to the center of gravity of the load.

Step 2   Repeat Step 1 for Crane Two.

Minimum net capacity of Crane Two:
90,500 lb − 500 lb (beam clamp) = 90,000 lb

Step 3   Determine the total net capacity of the cranes.

Total minimum net capacity cranes:
155,500 lb + 90,000 lb = 245,500 lb

Step 4   Verify that the load's weight is less than 75 percent of the sum of the two cranes' minimum net capacities.

$$\text{Percent capacity} = \frac{\text{Load weight}}{\text{Total net capacity}} \times 100$$

$$\text{Percent capacity} = \frac{180{,}000 \text{ lb}}{245{,}500} \times 100$$

$$\text{Percent capacity} = 0.733 \times 100 = 73.3\%$$

Figure 14 A two-crane coordinated heavy lift.

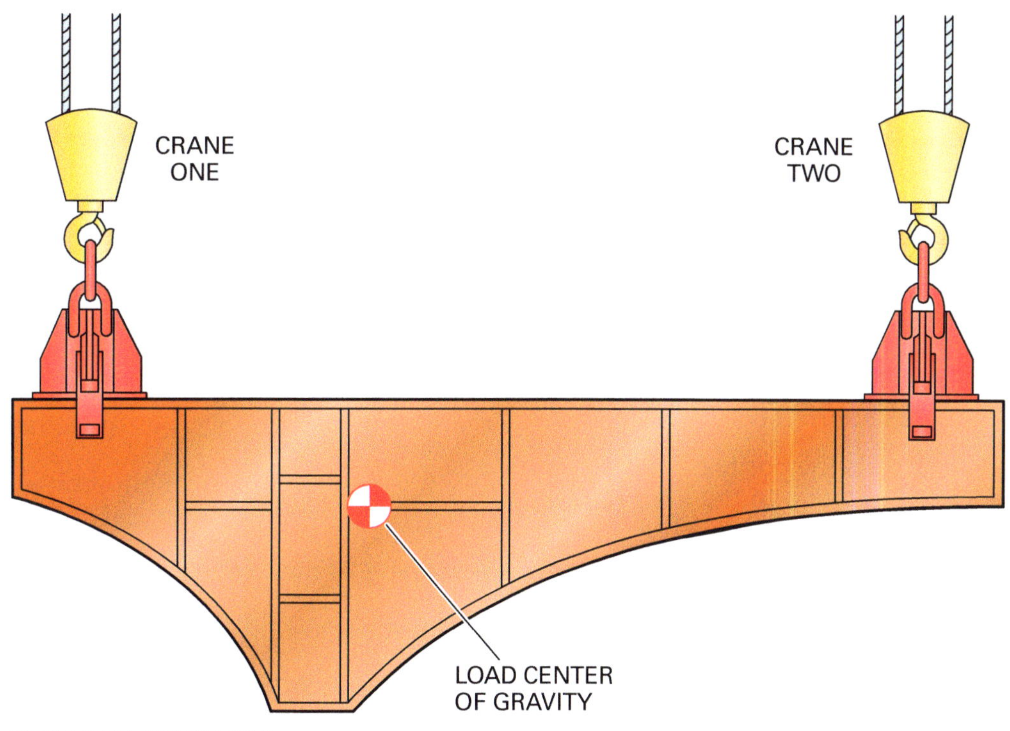

Figure 15 A dual-crane lift of a large structural member.

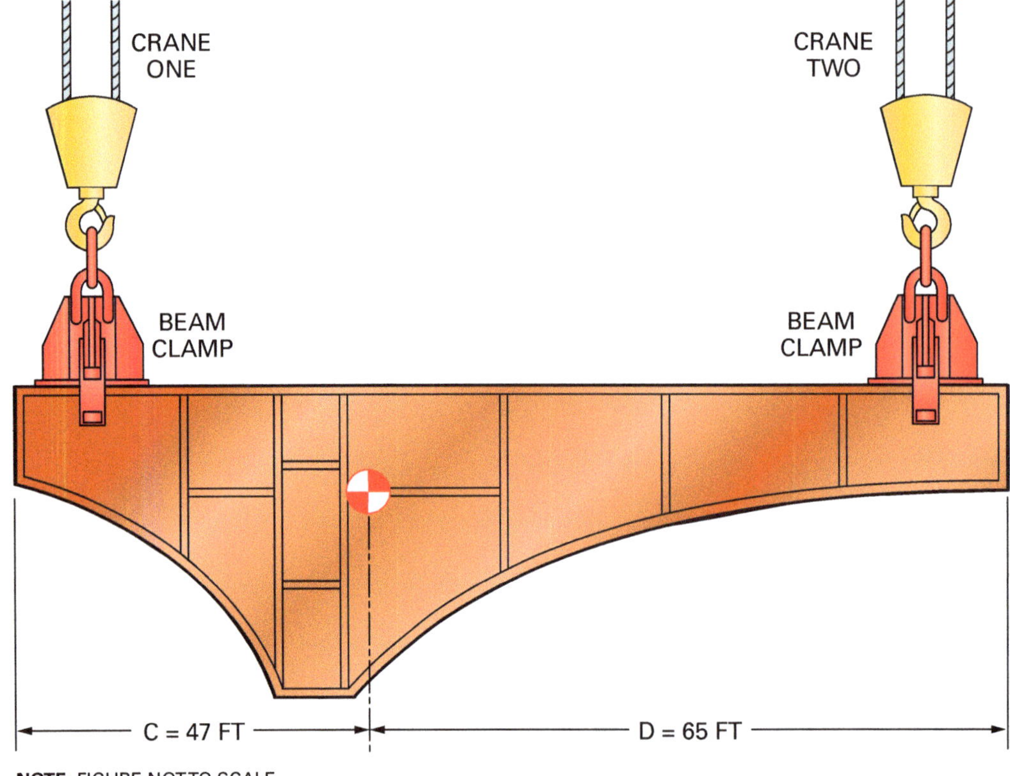

Figure 16 Location of the load's center of gravity.

Step 5 Locate the load's CG from the ends of the load, as shown in *Figure 16*. Obtain this information from builder's plans.

Step 6 Calculate the position of each crane's load block relative to the load's CG by using the following formula:

$$\text{Net capacity of Crane One} \times A = \text{Net capacity of Crane Two} \times B$$

Where:
A = distance of load block of Crane One from load CG
B = hook-block distance from load's CG for Crane Two

Step 7 Choose a trial length for A that will be less than the length of C in *Figure 16*. Start with A = 45 ft, for example (*Figure 17*). Then calculate B:

$$B = \frac{\text{Crane One net capacity}}{\text{Crane Two net capacity}} \times A$$

$$B = \frac{155{,}500 \text{ lb}}{90{,}000 \text{ lb}} \times 45'$$

$$B = 1.73 \text{ lb} \times 45' = 77.85' \text{ (round up to 78')}$$

This configuration will not work because the load block position for B (78 ft) would be greater than dimension D (65 ft) and beyond the end of the right side of the load (see *Figure 18*). To obtain a shorter value of B, choose a shorter length for A, say, 36 ft (*Figure 19*):

$$B = \frac{\text{Crane One net capacity}}{\text{Crane Two net capacity}} \times A$$

$$B = \frac{155{,}500 \text{ lb}}{90{,}000 \text{ lb}} \times 36'$$

$$B = 1.73 \text{ lb} \times 36' = 62.28' \text{ (round up to 63')}$$

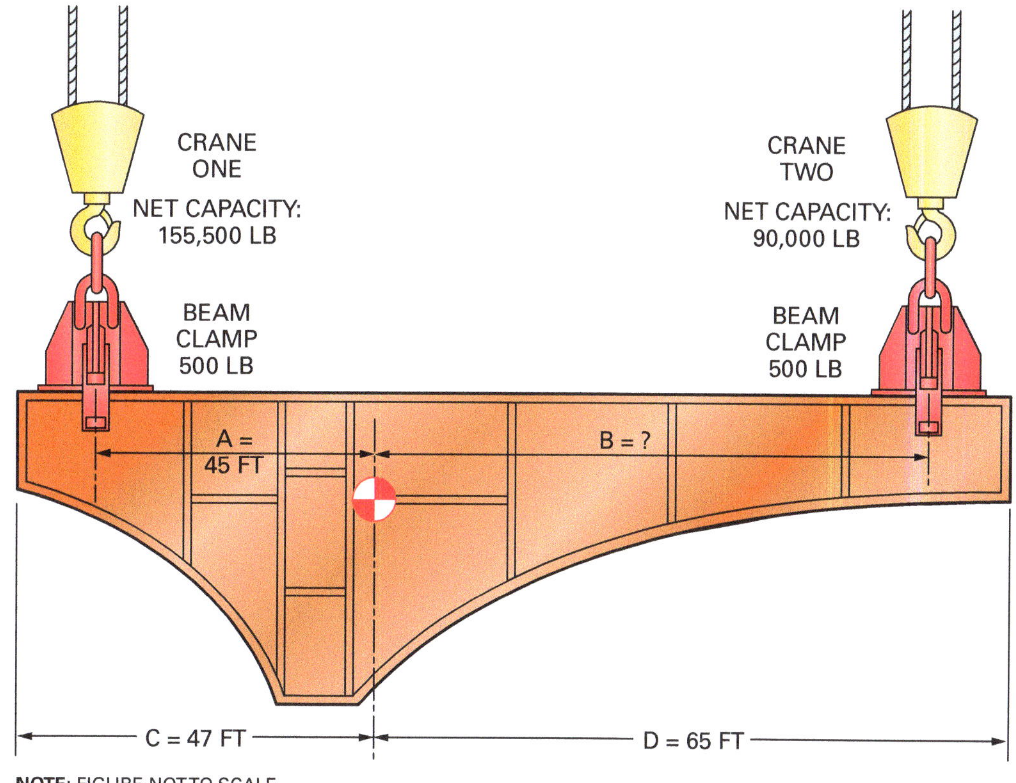

Figure 17 Calculating distance B, Trial 1.

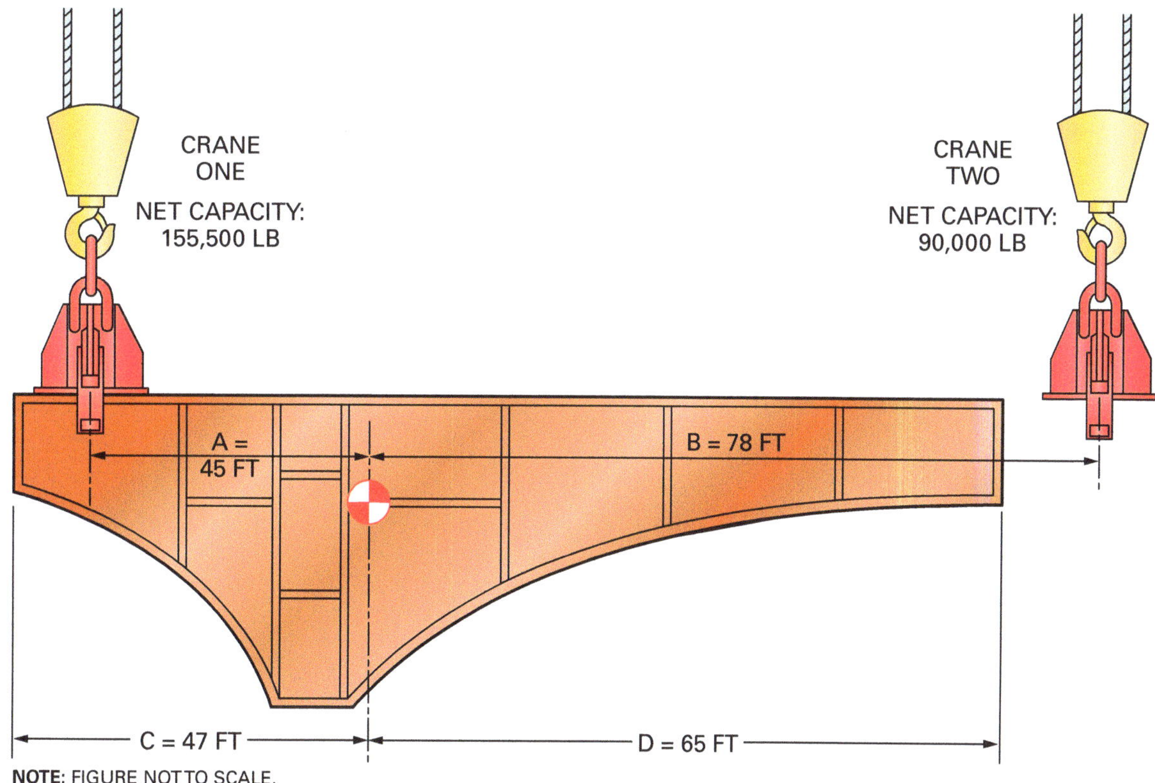

Figure 18 Calculating distance B, Trial 1 results.

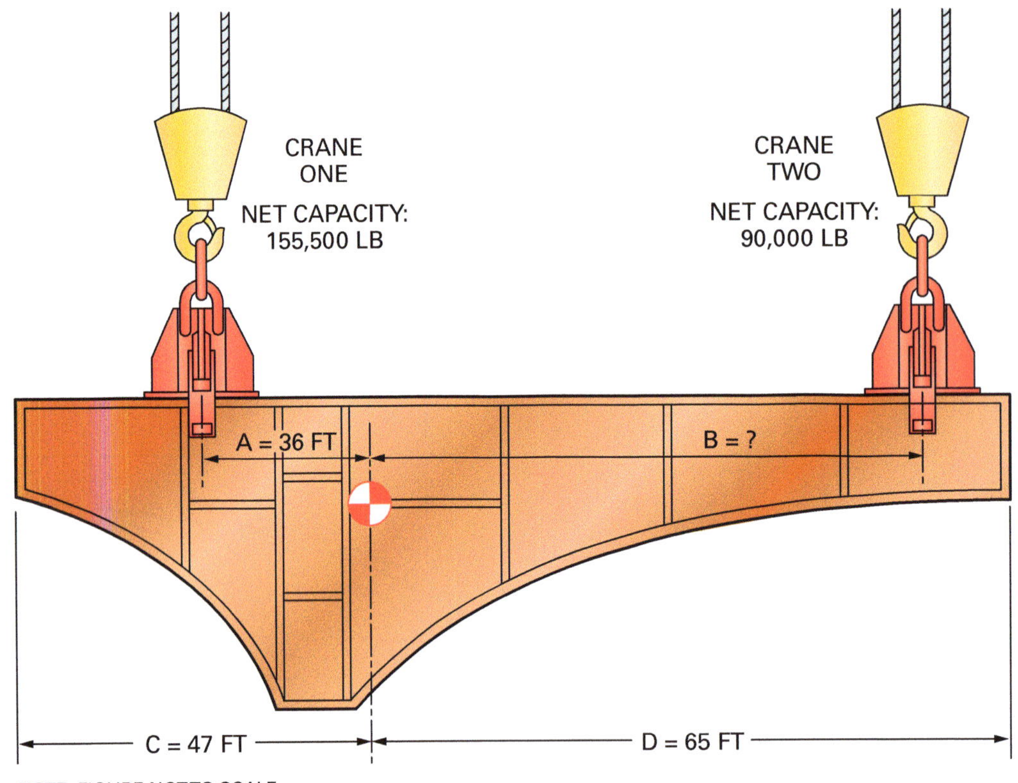

Figure 19 Calculating distance B, Trial 2.

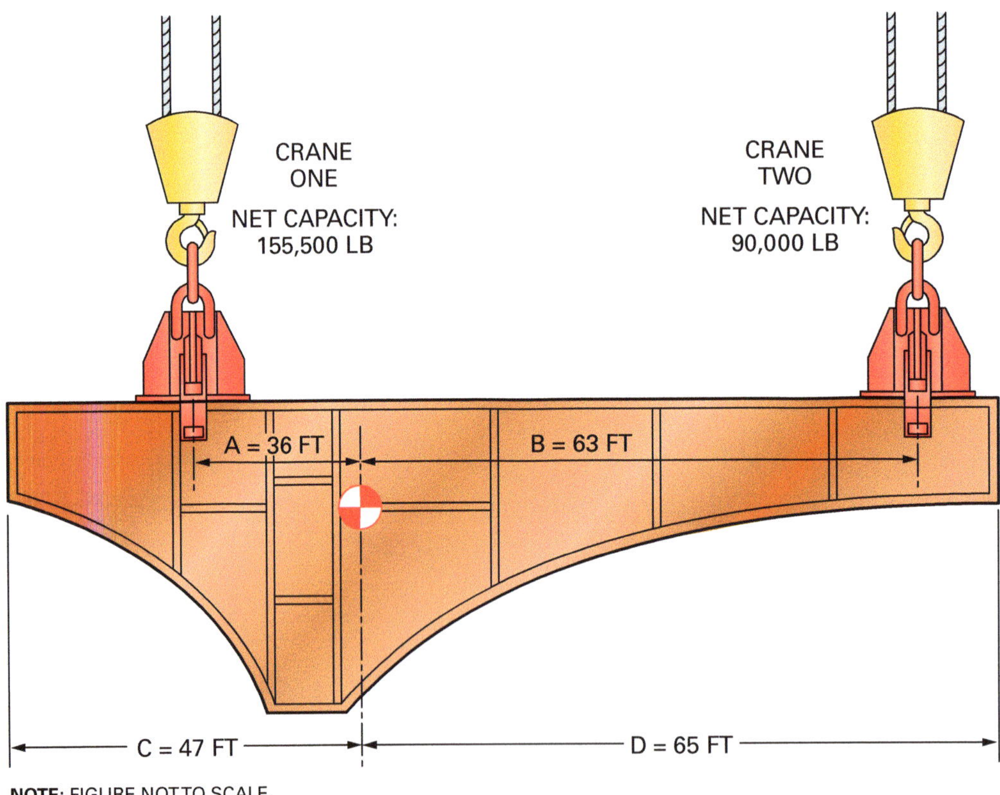

Figure 20 Calculating distance B, Trial 2 results.

NCCER – *Advanced Rigger*

Step 8 Determine the actual load that each crane will be carrying:

Crane One load =
   [B ÷ (A + B)] × load weight =
   [63' ÷ (36' + 63')] × 180,000 lb =
   0.64 × 180,000 lb = **115,200 lb**

(Less than 75% of min net capacity)

Crane Two load =
   [A ÷ (A + B)] × load weight =
   [36' ÷ (36' + 63')] × 180,000 lb =
   0.36 × 180,000 lb = **64,800 lb**

(Less than 75% of min net capacity)

Step 9 Since neither crane is loaded greater than 75 percent of its capacity, the two cranes can accomplish the lift as planned. Crane One's load block will be 36 feet from the load's CG and Crane Two's load block 63 feet from it.

1.3.1 Applying an Equalizer Beam

Multi-crane lifts can also use an equalizer beam (*Figure 21*) to distribute the weight in proportion to each crane's capacity. By adjusting the crane and/or load attachment points on the beam, it is possible to share the load weight between the cranes in any proportion desired.

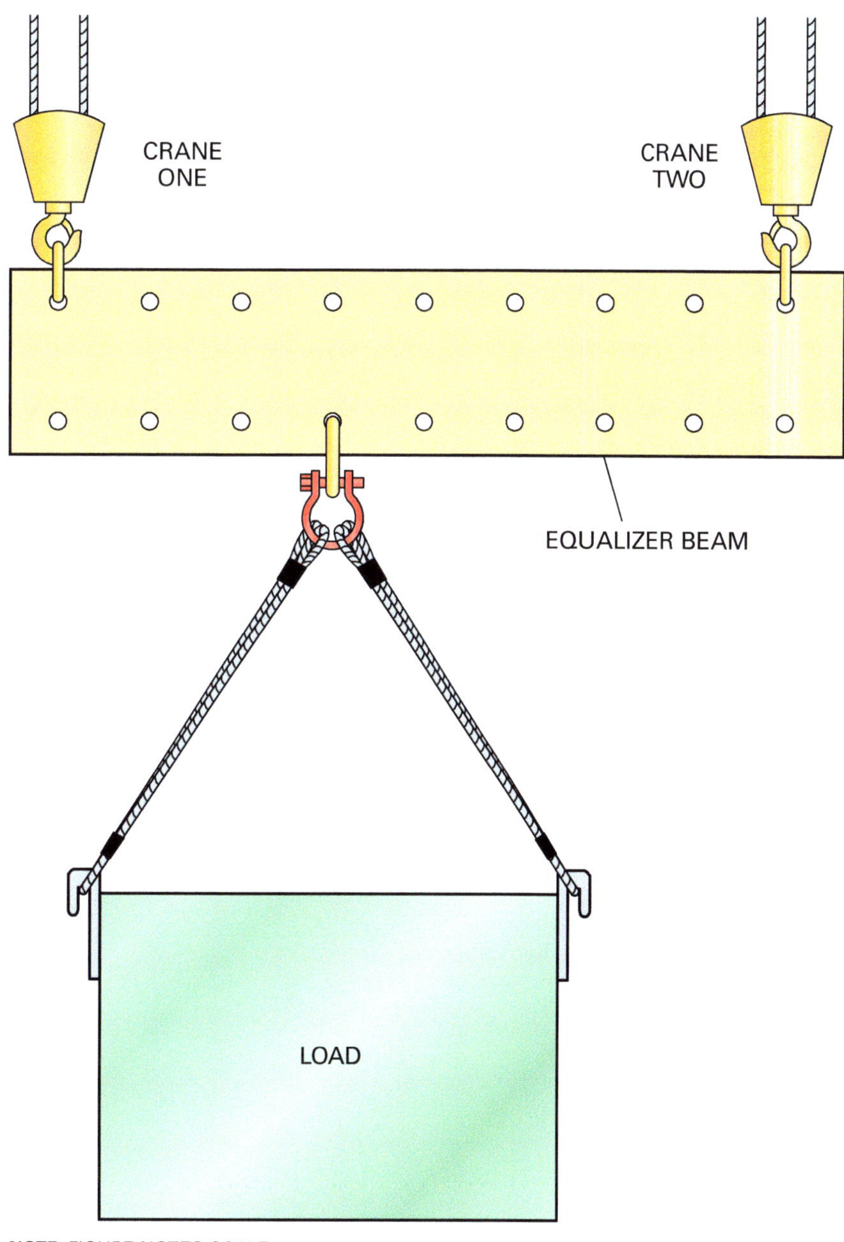

Figure 21 Dual-crane lift with equalizer beam.

When using an equalizer beam, measure the load radius of each crane to its pick-up point on the equalizer, not to the center of gravity of the load.

Using a process similar to the one described previously, the lift planner must determine the location on the equalizer beam where the load's hoist point will be, allowing each crane to share the load proportionate to its capacity.

For this example lift plan, the load is a 140,000-pound enclosure. Crane One has a minimum net capacity of 120,000 lb, and Crane Two's minimum net capacity is 80,000 lb. *Figure 22* shows the equalizer bar, the rigging, and the load scheduled for the dual-crane lift.

Step 1 Determine the total capacity of the cranes:

> *Minimum net capacities:*
> Crane One = 120,000 lb
> Crane Two = 80,000 lb
>
> *Total minimum net capacity:*
> Total capacity = Crane One + Crane Two
> Total capacity = 120,000 lb + 80,000 lb

This capacity is greater than the total suspended load weight of 143,000 lb (load + equalizer + rigging). Also, the load weight is less than 75 percent of the cranes' combined capacity (71.5%).

Step 2 Establish the load's hoist point:

For a simple beam problem like this example, the distance of one lifting force from the load's hoist point is proportional to the ratio of the *other* lifting force to the total weight of the load. The equations in the following steps illustrate this principle.

$$B = \frac{\text{Crane One net capacity}}{\text{Crane One capacity} + \text{Crane Two capacity}} \times A$$

$$B = \frac{80{,}000 \text{ lb}}{80{,}000 \text{ lb} + 120{,}000 \text{ lb}} \times 25'$$

$$B = \frac{80{,}000 \text{ lb}}{200{,}000 \text{ lb}} \times 25'$$

$$B = 0.40 \text{ lb} \times 25' = \mathbf{10'}$$

On the equalizer beam, if the distance $A = B + C$, then $C = A - B$:

$$C = A - B$$
$$C = 25' - 10'$$
$$C = 15'$$

The distance of Crane Two's load block from the hoist point is then 15 feet (shown in *Figure 23*).

Step 3 Determine the load that each crane will carry:

> Crane One load = (C ÷ A) × load weight
> Crane One load = (15' ÷ 25') × 143,000 lb
> Crane One load = 0.60 × 143,000 lb
> Crane One load = **85,800 lb**

This value is only 71.50 percent of Crane One's minimum net capacity of 120,000 lb.

> Crane Two load = (B ÷ A) × load weight
> Crane Two load = (10' ÷ 25') × 143,000 lb
> Crane Two load = 0.40 × 143,000 lb
> Crane Two load = **57,200 lb**

This value is only 71.5 percent of Crane Two's minimum net capacity of 80,000 lb. In this example, each crane is carrying the same percentage of their respective net capacity.

As you can see, even these calculations for a simple duo-crane vertical lift are not simple. A far more complex operation, such as slewing a load between cranes or rotating a long horizontal load into an upright position, can involve very complex dynamic calculations to identify the limiting configurations of the cranes during the lift. Riggers will not always be involved in this part of critical lift planning, but it is important for them to understand how lift planners perform these tasks.

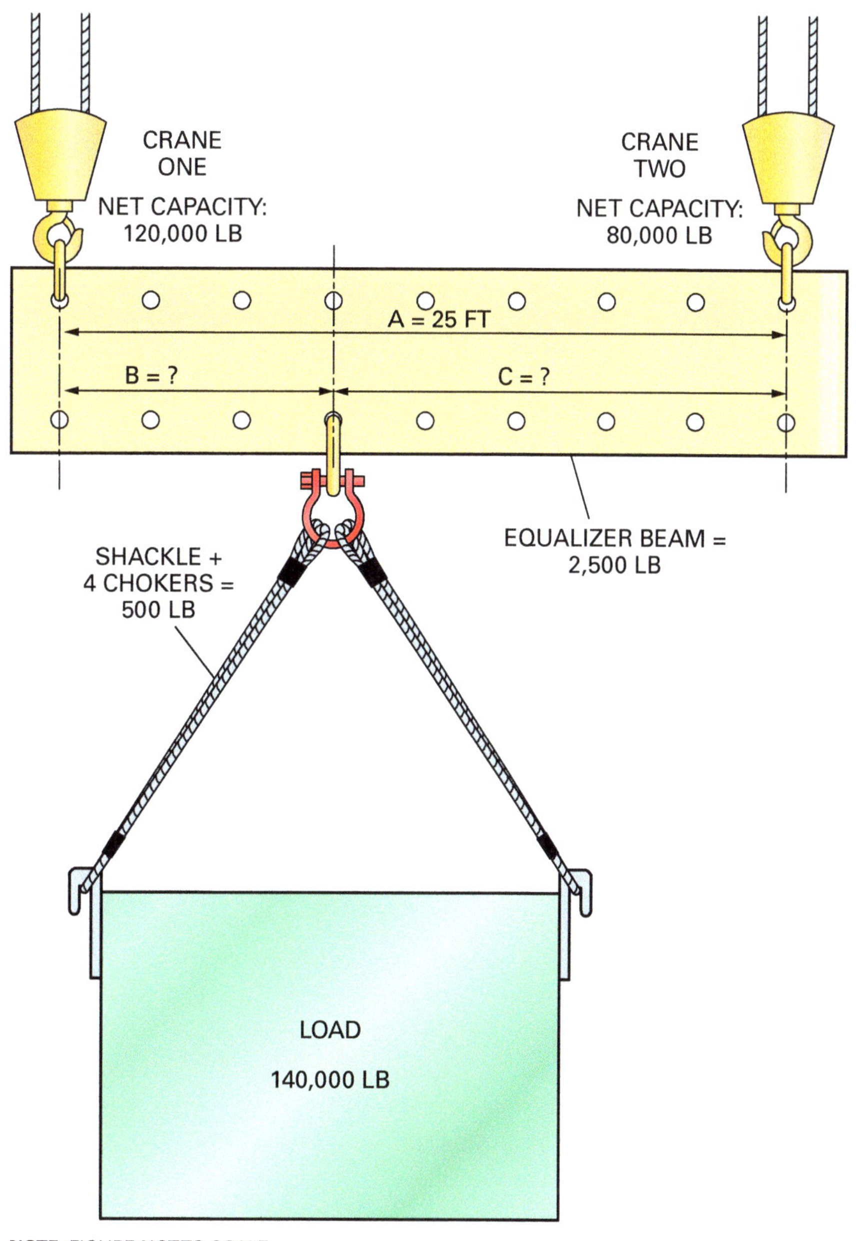

Figure 22 Dual-crane lift with equalizer beam—determining distances B and C.

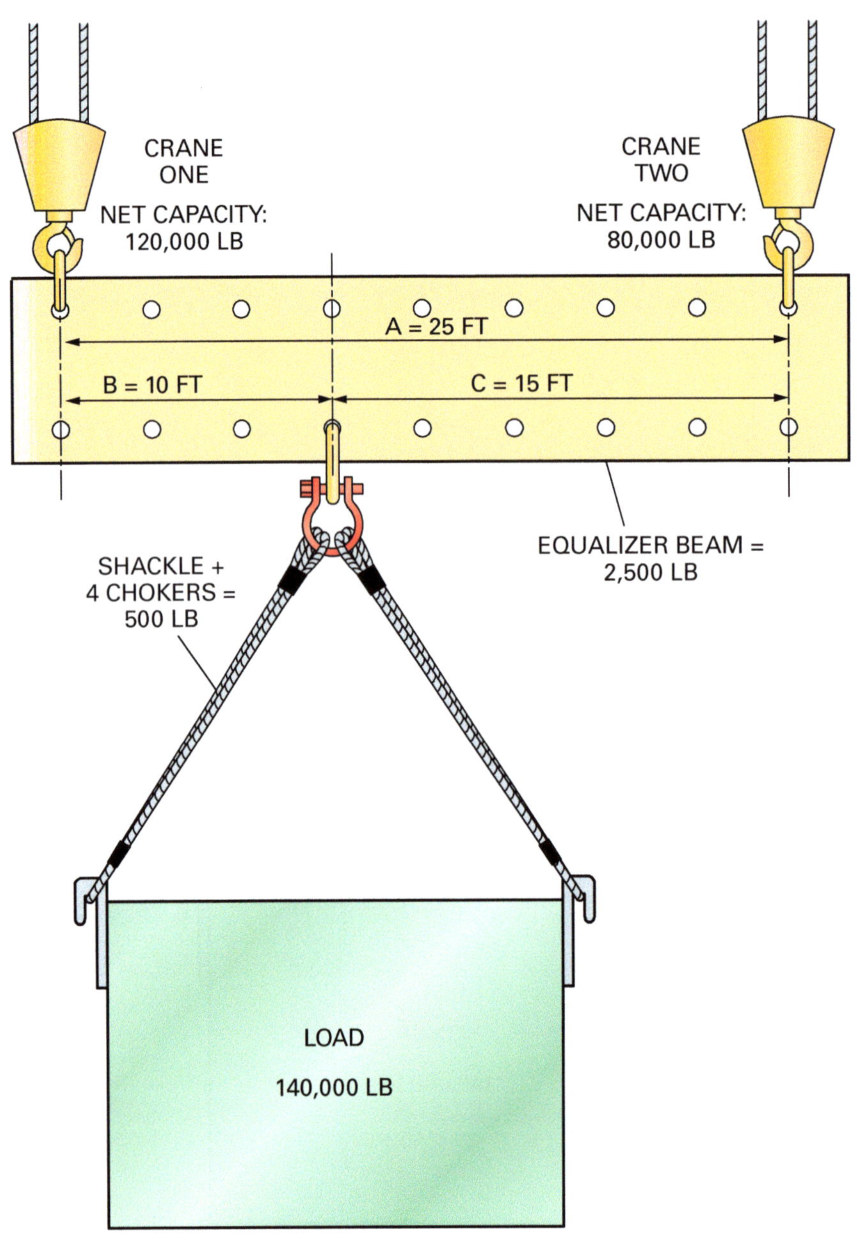

Figure 23 Calculated distances B and C.

Additional Resources

ASME Standard B30.5, *Mobile and Locomotive Cranes*. Current edition. New York, NY: American Society of Mechanical Engineers.

Crane Safety on Construction Sites, 1998. Task Committee on Crane Safety on Construction Sites. Reston, VA: ASCE.

29 *CFR* 1926, Subpart CC, *Cranes and Derricks in Construction*. **www.ecfr.gov**.

1.0.0 Section Review

1. Mechanical moment is the same as _____.

    a. speed
    b. force
    c. momentum
    d. leverage

2. Adding counterweights to the rear of the upperworks of a crane should have which of the listed effects?

    a. Moves the crane CG toward the counterweight
    b. Moves the crane CG away from the counterweight
    c. Mainly raises the overall crane CG
    d. Mainly lowers the overall crane CG

3. During a multi-crane lift, when the load's CG is off-center, simply having the cranes lift from each end would result in _____.

    a. one or both cranes exceeding their capacity
    b. one crane carrying more of the load than the other
    c. each crane carrying the same amount of weight
    d. damage to the load

2.0.0 SPECIALIZED EQUIPMENT USED IN HEAVY RIGGING

Objective

Identify and describe the use of special rigging equipment.

a. Describe the use and application of cribbing.
b. Explain how to determine line pull when using inclined planes to move equipment.
c. Identify and describe of spreader bars and equalizer beams.
d. Explain how to rig and handle reinforcing bar bundles.

Performance Task

1. Select the appropriate spreader bar or equalizer beam for a given load.

Trade Terms

Chicago boom: A load-hoisting device consisting of a single boom attached at its base to either a vertical post or beam in a larger structure, such as a building skeleton, or a freestanding pole stabilized by guy wires. Such booms are easily erected and suitable for repetitive hoisting operations at a fixed location.

Coefficient of friction (CF): A ratio that expresses a comparison between the force necessary to move an object over the surface of another material, and the pressure between the two materials.

Cribbing: A stack of heavy squared timbers laid in a crisscross pattern to temporarily support loads during movement and placement.

Inclined planes: Flat surfaces placed at a shallow angle that connect levels at different heights. A ramp is an example of an inclined plane.

Rebar: Shortened form of the term reinforcement bar. Used to identify individual rods, bundles, and field-fabricated meshes of rods that reinforce poured concrete structures.

Cranes are the most common hoisting equipment used in the rigging profession. However, riggers must also be familiar with other rigging methods, including the use of jacks, tuggers, rollers, grip hoists, and inclined planes. This section describes the use of cribbing and inclined planes for heavy rigging, and it finishes with a practical application of key rigging principles.

2.1.0 Cribbing

Cribbing is used to temporarily support a heavy weight or raise an object to a height at which simple blocking would be unstable. Riggers build cribbing by stacking squared-off hardwood timbers in alternating crossed tiers (*Figure 24*).

When placing cribbing, it is important to consider the stability of the foundation. The timbers must rest firmly and evenly on the ground. If the ground needs leveling, dig away and level any high spots rather than filling in low areas with loose material.

Cribbing helps support the load when raising or lowering it in successive stages. To raise a load, place the jacks under it on blocking on the ground. Raise the load with the jacks to their maximum height, then place one or more layers of cribbing under the load. Lower the jacks and continue the process by adding blocking under them so that the rams almost touch the load in their lowered position. Several different thicknesses of cribbing may be required, depending on the jacks used and their range of ram extension. Repeat the jacking and cribbing process until the load is at the desired height. Riggers can also lower a load in steps by reversing this process.

The nominal size of the hardwood lumber used in both blocking and cribbing usually starts at 4×4, but may be as large as 14×14, depending on the application. Every rigger should understand the following rules concerning the working load limit (WLL) of wooden beams:

- The greater the depth of a timber, the greater the WLL of the beam. As a general rule, doubling the depth of a beam increases the WLL by four.
- Two similar beams placed together are twice as strong as a single beam. Secure the ends with nails or dowels to stabilize them and ensure their strength.

Figure 24 Alternating crossed tiers of timber.

- The more uniformly a load is distributed, the greater the load that can be carried by a beam of a given size, material, and length. A load concentrated at the center of a beam reduces the WLL of that beam by one-half.

2.2.0 Inclined Planes

An inclined plane is a ramp that slopes gradually upward or downward. Riggers use inclined planes to raise or lower an object from one level to another. Less force is required to move an object up an inclined plane than to lift it vertically off the ground. A rigger can use a sketch and some simple arithmetic to estimate the amount of pull required to move a load up an inclined plane.

Estimating the line pull for an inclined plane requires knowing or measuring the following quantities, as shown in *Figure 25*:

- Weight of the load – W
- Length of the ramp – L
- Height between the two surfaces – H
- Horizontal run of the ramp – R

Riggers also need to identify the materials of the load and the ramp in order to determine how much friction will occur when moving the load. Friction opposes the motion between two objects in contact. Therefore, the force to overcome friction will add to the line pull needed to move the load up a frictionless ramp.

Friction results from attractive forces between substances at the molecular level. These forces relate to the kinds of materials, how closely they touch, and the amount of weight, force, or pressure there is between the surfaces. For any two materials and their surface characteristics, the friction is proportional to the force pressing them together (the force perpendicular to their surfaces—W_{Perp}). Scientists call the proportionality constant the co-efficient of friction (CF). The CF has values between zero and one for non-adhesive materials.

The formula for calculating the friction between two materials is:

$$\text{Friction} = \text{CF} \times W_{Perp}$$

Practically speaking, if an object of one material rests on the level surface of another, and they have a CF of 0.5, you would need to exert half the object's weight (W) sideways to overcome the friction and get it moving. Remember that on a horizontal surface $W = W_{Perp}$. In other words, when the object is resting on a level surface, it's actual weight and the weight it imposes on that surface is the same.

Engineers determine coefficients of friction experimentally, and references list them for many common pairs of materials. *Table 2* is a short list of CFs that can apply to rigging situations. Interestingly, the amount of area in contact doesn't affect friction; friction is affected only by the types of materials.

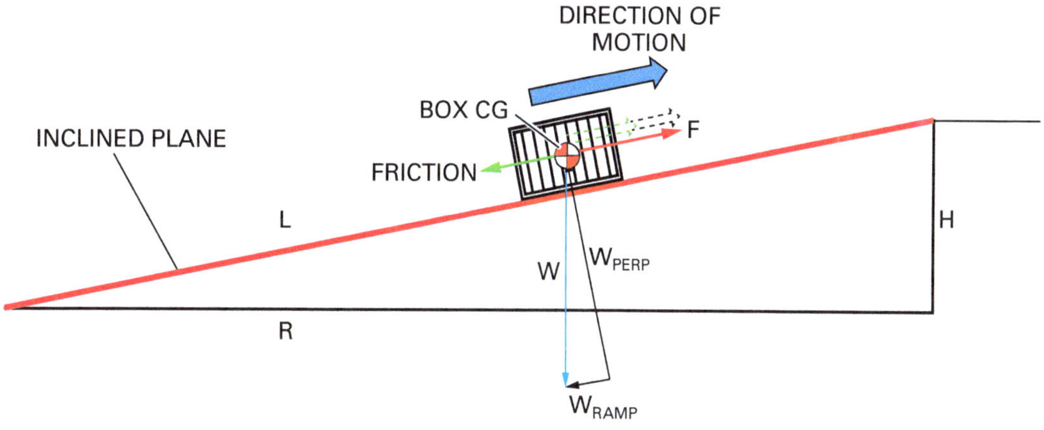

Figure 25 Line pull for a load on an inclined plane.

Table 2 Examples of Coefficients of Friction

Materials		Coefficient of Friction (Static)
Concrete	Concrete	0.65
Concrete	Metal	0.60
Concrete	Wood	0.45
Concrete	Rubber	0.90–1.0
Steel	Steel	0.74
Steel	Steel (lubricated)	0.15
Steel	Aluminum	0.60
Steel	Cast iron	0.25
Wood	Wood (species-dependent)	0.25–0.50
Wood	Metal	0.30
Wood	Manila rope	0.40
Teflon™ (PTFE)	Teflon™ (PTFE)	0.04

In addition to friction, the minimum line pull required is proportional to the weight of the load and the dimensions of the inclined plane. This is because some of the load's weight tends to move the load down the ramp. If the ramp were frictionless, it would slide to the bottom due to this fraction of its weight. To simplify, assume that all the forces involved act on the CG of the load. In *Figure 25*, the weight triangle's symbols mean the following:

- W is the load's dead weight.
- W_{Perp} is the portion of the load's weight perpendicular to the ramp's surface (required for calculating friction).
- W_{Ramp} is the portion of the load's weight tending to move the load down the ramp.

Notice that the weight triangle (W-W_{Perp}-W_{Ramp}) is geometrically similar to the inclined plane triangle (L-R-H). This means that the weight of the load (W) is proportional to W_{Perp} and W_{Ramp} in the same way that the length of the ramp (L) is proportional to R and H. These relationships exist because of the rules for similar triangles.

The minimum line pull to move the load up the ramp (F) is the sum of the forces required to overcome friction and to oppose the weight of the load tending to move it down the ramp. These forces are indicated by the dashed arrows above the force arrow in the figure.

Calculate the force required to overcome friction using the formula. First, you must find the perpendicular fraction (W_{Perp}) of the load's weight. Use the following proportion:

$$\frac{W_{Perp}}{W} = \frac{R}{L}$$

Solve for W_{Perp} by multiplying both sides by W:

$$W_{Perp} = W \times \frac{R}{L}$$

Substitute into the friction equation above:

$$\text{Friction} = CF \times \left(W \times \frac{R}{L}\right)$$

Calculate the part of the load's weight tending to move it down the ramp (W_{Ramp}). Use the following proportion:

$$\frac{W_{Ramp}}{W} = \frac{H}{L}$$

Solve for W_{Ramp} by multiplying both sides of the equation by W:

$$W_{Ramp} = W \times \frac{H}{L}$$

The combined formula to calculate the minimum line pull is:

$$F = \text{friction} = W_{Ramp}$$

Substituting:

$$F = \left[CF \times \left(W \times \frac{R}{L}\right)\right] + \left[W \times \frac{H}{L}\right]$$

> **NOTE**
>
> The inclined plane minimum line pull formula looks complicated, but is commonly used. Be sure to use the proper arithmetic order of operations, working from the innermost parentheses outward. Also, make sure that you substitute the correct values for their symbols in the formula.

Example:

Assume the steel box in *Figure 25* weighs 500 pounds (W). The aluminum ramp length is 10 feet (L). The height to gain is 2 feet (H) and the run of the ramp is 9.8 feet (R). The coefficient of friction between the steel box and the aluminum ramp from *Table 2* is 0.60. Calculate the minimum line pull (F) to move the box up the ramp.

$$F = \left[CF \times \left(W \times \frac{R}{L} \right) \right] + \left[W \times \frac{H}{L} \right]$$

$$F = \left[0.60 \times \left(500 \text{ lb} \times \frac{9.8 \text{ ft}}{10 \text{ ft}} \right) \right] + \left[500 \text{ lb} \times \frac{2 \text{ ft}}{10 \text{ ft}} \right]$$

$$F = [0.60 \times (500 \text{ lb} \times 0.98)] + [500 \text{ lb} \times 0.20]$$

$$F = [0.60 \times 490] + 100 \text{ lb}$$

$$F = 294 \text{ lb} + 100 \text{ lb}$$

$$F = 394 \text{ lb}$$

The result gives you the minimum force needed to get the load moving. In reality, once the load starts to move, the friction decreases, requiring less pull to move the load. This is because the coefficient of friction for a moving contact between two materials (called *coefficient of kinetic friction*) is smaller than for the stationary CF for the same materials (*coefficient of static friction*).

2.3.0 Lifting Beams

There are times when slings alone are not appropriate for a given lift, especially when rigging large, unwieldy loads that might tilt or slip. Also, some loads are not strong enough to handle lifting from a single point, even if they are balanced. Riggers use beams to distribute the weight of a load equally. While an engineer will determine the beam configuration in most cases, the rigger must be able to calculate the sling angle appropriate to the load and the length of sling needed to achieve the desired angle.

There are two basic types of lifting beams:

- Below-the-hook attachments that directly suspend the load from points anywhere along their lengths
- Spreader beams, which mainly exert forces at the ends to help position slings and bridles attached to the load to maximize their WLL

True lifting beams of the first type must handle bending and shear forces imposed by the load. Spreader beams mainly deal with compression forces oriented along their lengths generated by the slings attached to them at their ends. Both types may have attachment points for suspending the load. All lifting beams are required to have their WLL and weight stenciled on the beam.

2.3.1 Spreader Beams

Riggers use spreader beams to support long, flexible, or unbalanced loads, such as packs of roof sheeting or bundles of reinforcing bar (rebar). If used correctly, spreader beams help to protect the load from tipping, sliding, or bending. Spreader beams also help to maximize sling angles at the load, as well as the tendency of bridles to crush long loads.

Figure 26 shows several spreader beams in use to lift a large piping module, including the supporting structure around it, from a single hook. Note the complexity of the arrangement, but also consider that the load placed on each sling and point of connection is calculated in advance. Such complex arrangements are not created "on the fly." With careful work by lift planners, the entire arrangement is made to work on paper or in software before the components ever leave the building. Notice also the dual load blocks and the many parts of line reeved.

Riggers need to be aware of the attachment points on the load before selecting rigging equipment. Shoulder-less eyebolts may only be used for vertical lifts or hitches; pulling at an angle is not allowed. The same applies to most welded lifting lugs. In general, lifting lugs withstand more force when lifted from directly above than they will when lifted from an angle. Riggers often use spreader beams to avoid lifting such hardware at an angle. The use of spreader beams provides better load control and more even distribution of the load for certain types of lifts. *Figure 27* shows a lift using multiple spreader beams with a four-leg bridle.

Engineers normally design spreader beams to meet a specific standard, and the beams are proof-tested before riggers may use them to make a lift. Spreader beams for offshore applications require testing to a much higher overload due to the inherent higher dynamic forces resulting from snatching loads in rough sea conditions.

Figure 26 Spreader beam application.

Figure 27 Multiple spreader beam application.

2.3.2 Adjustable Lifting Beams

Adjustable beams are popular because riggers can use a single beam for many different load sizes. Before the availability of adjustable beams, it was often necessary for riggers to keep a variety of beam sizes on hand to deal with the many loads they might encounter. Telescopic beams are a good example of adjustable lifting beams. These beams have a main body constructed as a hollow box-beam with one or more hoist attachment points. The end sections with the load attachments slide in or out from the ends. Hardware fastens them in place at the required spread.

Some adjustable beams, such as the ones shown in *Figure 28*, are made from two end caps attached to a length of pipe. The diameter, wall thickness, and length of the pipe determine the beam's lifting capacity. Rigging firms can use the same caps with different lengths of pipe, reducing costs as well as inventory. Some models are modular, where shorter pipe sections bolt together with flanges to make longer beams. Common pipe diameters for this purpose are 6" and 8", but larger ones are fabricated as needed. Some pipe beams may be 48" in diameter and up to 40 feet in length. The beam manufacturer provides a capacity chart that relates the length and size of the pipe to load capacity.

Other types of adjustable beams have a fixed length, but provide multiple or movable connection points along the length of the beam, as shown in *Figure 29*. These adjustable beams are known as *equalizer beams*. In some models the lifting points, the load attachment points, or both are movable.

The primary function of an equalizer beam is to rig an asymmetrical load so that its CG is directly under the hook, equalizing the rotational forces on the load. They are also useful for distributing the weight of a load between two cranes when making tandem crane lifts. Crane hoists attach to the ends of the beam while the load is suspended from beneath it.

Engineers usually design equalizer beams for special lifts. As with other below-the-hook devices, the beam must bear a tag or label showing the WLL, beam weight, and other information required by *ASME Standard B30.20*. The lift chains

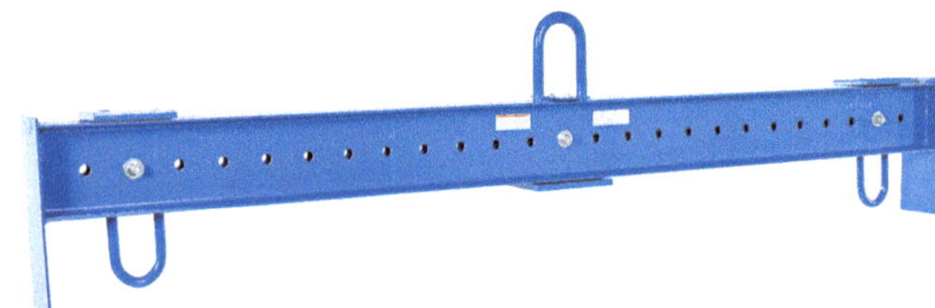
Figure 29 Equalizer beam.

(A) END CAPS

(B) INTERCONNECTING PIPE

Figure 28 Adjustable end cap beam.

or wire rope slings are usually captive to and tested with the beam. The beam may be stored on a purpose-built rack when not in use.

2.4.0 Rigging Rebar Bundles

This section presents a common rigging operation that incorporates many of the rigging concepts described previously. A rebar bundle (*Figure 30*) makes an interesting load to rig for lifting. The individual bars themselves can vary in diameter and length. The bundles are heavy, but flexible in long lengths. Rebar usually arrives at the jobsite on flatbed trucks. If a spur track is available or adjacent to the area, the work site may receive shipments by railcar.

2.4.1 Unloading Procedure

Accompanying each shipment is a bill of lading, which lists the materials contained in the shipment. As the material is unloaded, the responsible person must check the bundles against the shipping list to account for all materials. Receiving personnel should record the weight and other necessary information for each shipment. The trucks should be unloaded promptly so that the job site can release them, preventing additional charges.

After receipt, workers may deliver the rebar to a storage area, where a foreman directs unloading. If rebar delivery is on a schedule to meet the daily placement requirements, the loads are delivered to the points of placement or to the crane so that the bundles can be hoisted to the placement area. Make sure that the lifting equipment

Figure 30 Rebar bundles.

has the necessary lifting capacity and reach for handling the load and that materials or rigging accessories are available to pad any sharp edges.

> If a load requires more than two people to carry it, use lifting equipment to prevent personal injury.

Ensure that the area where the load is set down is strong enough to support the weight. Supervisors should establish safe load limits before unloading. Place timbers on the ground to support the rebar bundles and keep them free from mud and other jobsite debris. Select blocking and timbers with care. Workers should use hardwood that is large enough to allow safely blocking or cribbing the materials. Avoid using worn blocking that has rounded corners or shows signs of dry rot.

2.4.2 Rigging and Handling

Workers use power equipment to handle rebar for raising, lowering, placement, and bridging obstacles. The type of hoisting equipment depends on the size of the job, site congestion, and structure height. The most commonly used hoists are truck and telescopic cranes, and the Chicago boom. When hoisting bundles of bars measuring 30 feet or longer, always use a spreader beam or lifting beam to prevent bending of the bars. The beam should be at least half the length of the bars. *Figure 31* shows the use of a lifting beam to handle rebar.

When hoisting bundles in congested or confined areas, always attach a tagline to one or both ends of the load to guide it during the operation. Taglines are commonly made of fiber rope. Safe hoisting requires a skilled operator taking instructions from an authorized signal person using standard hand signals.

Riggers should never hook a hoist line to the wire wrapping used to tie the bundle of rebar together. This wire is not strong enough to carry the load. Use slings made of wire rope or chain with the proper capacity for the weight of the bundle. The sizes of hooks and other rigging hardware must have the necessary load capacity as well.

Ensure that the slings are strong enough to lift the load. The stress on each sling depends on the number of slings, the angle of the sling, and the total load including the rigging. When using two choker hitches to lift bundles, make the chokes in the same direction. If the chokes wrap from opposite sides, the bundle will twist when hoisted. *Figure 32* shows the proper way to use two slings to lift a bundle.

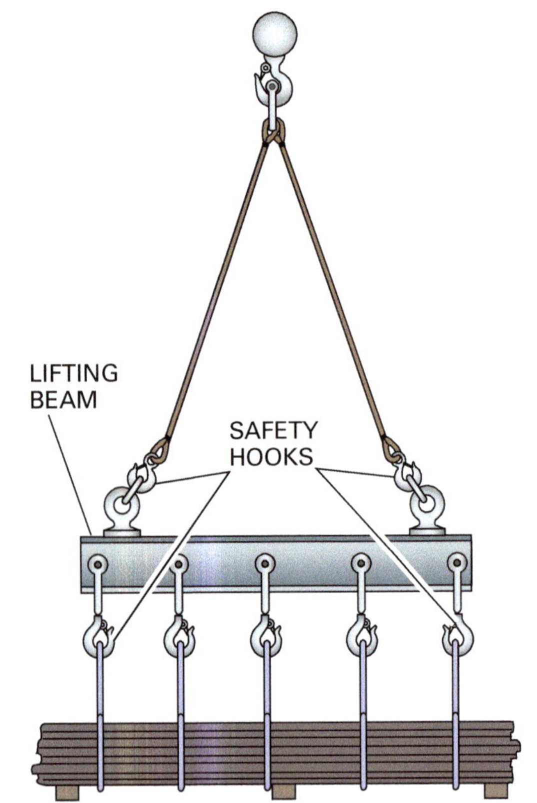

Figure 31 Use of a lifting beam to rig rebar.

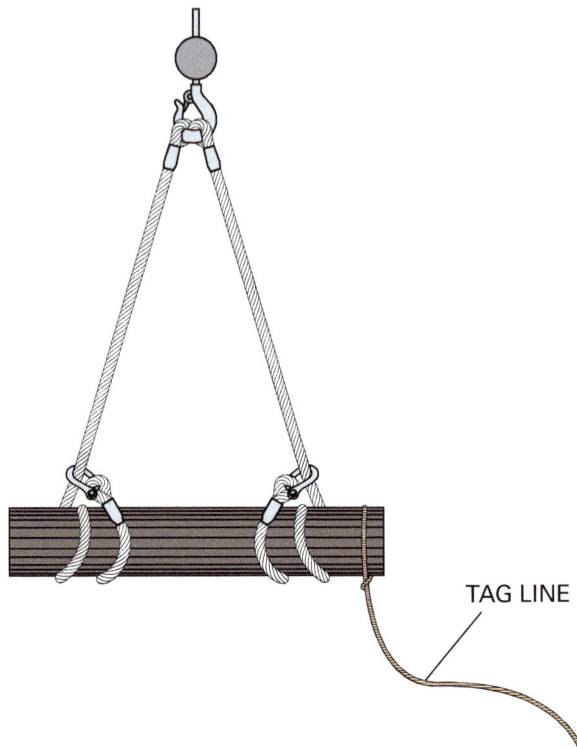

Figure 32 Using double-wrapped choker hitches to lift a bundle.

Additional Resources

ASME Standard B30.20, Below-the-Hook Lifting Devices. Current edition. New York, NY: American Society of Mechanical Engineers.

ASME Standard P30.1, Planning for Load Handling Activities, Current edition. New York: The American Society of Mechanical Engineers.

Bob's Rigging and Crane Handbook, Latest edition. Leawood, KS: Pellow Engineering Services.

Rigging Handbook, 4th Edition. 2012. Jerry A. Klinke. Stevensville, MI: ACRA Enterprises, Inc.

Willy's Signal Person & Master Rigger Handbook, First Edition. 2013. Ted Blanton Sr, Robert O'Leary, Joe Crispell, and Ted Blanton Jr. Lake Mary, FL: NorAm Productions, Inc.

2.0.0   Section Review

1. Cribbing is useful for _____.

    a. protectively surrounding a load like a crib
    b. supporting a heavy load as riggers raise or lower it into position
    c. moving a load from a lower to higher level using less force than required to lifting it vertically
    d. stabilizing a load during a crane lift

2. Calculate the force needed to overcome the friction between a 100-pound metal box resting on a concrete sidewalk. Refer to *Table 2*.

    a. 60 lb
    b. 67 lb
    c. 160 lb
    d. 167 lb

3. Which of the following statements about spreader beams is *correct*?

    a. They typically have multiple load attachment points along their lengths.
    b. They help decrease the sling angles of slings supporting the load.
    c. They are appropriate for lifting mainly boxy, rigid loads.
    d. They are appropriate for lifting long, flexible, or unbalanced loads.

4. When receiving and unloading rebar, which of the listed actions is *not* appropriate?

    a. Place the rebar on blocking on firm ground.
    b. Unload the rebar at the placement site.
    c. Check the materials against the shipping list after releasing the trucks so they don't charge for delays.
    d. Make sure that load handling equipment has sufficient rating for the loads they will carry.

SUMMARY

Chemical plants, power plants, paper mills, oil refineries, and other industrial operations require the hoisting or movement of immense equipment, machines, and structures. Such operations may require the use of more than one crane. In many situations, it is necessary to move a piece of equipment or a structure horizontally to position it for hoisting. Riggers use specialized equipment and devices for these tasks. The rigger needs to know which type of device to use and must be able to select the device with the correct capacity for the application.

Workers must have specialized knowledge to plan and implement complex rigging requirements. Many factors can affect a lifting operation, and each factor can cause a load to act in a different manner. It is up to the crane operator and rigging specialist to understand how to prevent conditions that might cause a crane to tip. For example, it is important to know how the tipping axis can change simply by rotating the crane to an over-the-side quadrant. All workers involved with planning a lift must consider such factors during development of the lift plan and during the rigging operation itself.

Although engineers may design the lift and even the hardware used, the rigging specialist plays an important role in planning and carrying out the lift. At every step in the process, the rigger must ensure that the operation is performed safely and the correct equipment and methods are used.

NCCER – *Advanced Rigger*

1. The term *dynamics* as used in mobile crane operations refers to ______.

   a. generating power
   b. the results of applied forces
   c. explosives
   d. interpersonal relationships

2. For most truck-bodied cranes on outriggers, the most stable quadrant of operation is usu-ally ______.

   a. over-the-front
   b. over-the-side
   c. over-the-rear
   d. throughout the full 360-degree arc

3. Impact or dynamic loading directly relates to changes in ______.

   a. the weight of the load
   b. the weight of the crane
   c. the load moment arm
   d. the motion of the load or boom

4. Which of the following statements about dual-crane lifts is *correct*?

   a. Both operators must keep their hoist lines vertical at all times.
   b. Since there are two, using 80 percent of each crane's capacity is common.
   c. There should always be at least three sig-nal persons involved to ensure clarity.
   d. The two cranes must always be the same basic model and size to simplify engineering.

5. Which of the following statements about cribbing is *correct*?

   a. Riggers use cribbing when they want to raise a load into position in a single step.
   b. Riggers cut cribbing to the proper length and place them on end under the load.
   c. Blocking is preferable to cribbing when load support is a concern.
   d. Riggers build up cribbing in crossed tiers of timbers.

6. An inclined plane is useful for ______.

   a. raising or lowering a load from one level to another
   b. moving a load across a utility trench at the work site
   c. creating a mechanical moment to turn a load on a hoist line
   d. bracing a load at an angle

7. A rigger needs to move a 500-pound metal shipping container on wooden skids up a metal ramp to a dock positioned 4 feet above the pavement. The ramp is 20 feet long. The lower end of the ramp is 19.6 feet from the base of the dock. Estimate the minimum line pull required to move the container up the ramp. Refer to *Figure RQ01*, shown here, and *Table 2*.

   a. 1,200 lb
   b. 516 lb
   c. 247 lb
   d. 100 lb

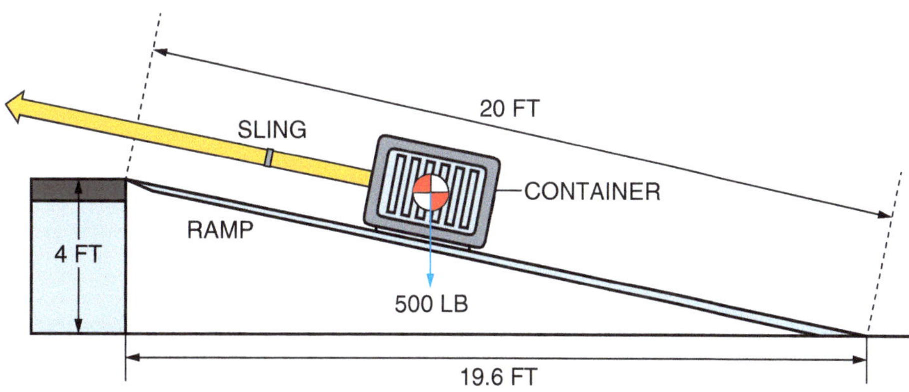

Figure RQ01

8. Mechanically speaking, the main job of a spreader beam is to _______.

   a. handle compression forces generated by slings attached at their ends
   b. handle bending and shear forces when lifting a load
   c. decrease sling angles to maximize bridle WLL
   d. eliminate the need for load attachment points

9. Rigging a bundle of rebar can be challenging because _______.

   a. it is expensive and easily damaged
   b. it has sharp edges
   c. rebar must be lifted one piece at a time
   d. the bundles are heavy and flexible in long lengths

10. Which of the statements about handling rebar is *correct*?

   a. Lifting rebar by the wires securing them is allowed because they are designed for that purpose.
   b. Use taglines to guide the lifting of rebar in congested and confined areas.
   c. Dual choker hitches should be wrapped in opposite directions around rebar bundles to keep them from twisting.
   d. Acute sling angles when lifting long rebar bundles will minimize bending the rebar.

Trade Terms Introduced in This Module

Backward stability: In crane operations, the measure of a crane steadiness in relation to the tipping axis on the opposite side of the crane from the boom.

Center of gravity (CG): The point where one can assume all of an object's mass, and therefore its weight, is concentrated. The concept is useful for determining stability, a load's balance point, and leverage.

Chicago boom: A load-hoisting device consisting of a single boom attached at its base to either a vertical post or beam in a larger structure, such as a building skeleton, or a freestanding pole stabilized by guy wires. Such booms are easily erected and suitable for repetitive hoisting operations at a fixed location.

Coefficient of friction (CF): A ratio that expresses a comparison between the force necessary to move an object over the surface of another material, and the pressure between the two materials.

Combined CG: The sum of the crane's and load's CGs at a position along an imaginary line connecting the two CGs.

Cribbing: A stack of heavy squared timbers laid in a crisscross pattern to temporarily support loads during movement and placement.

Dynamics: The response of objects to the application of forces, moments, or torques.

Forward stability: In crane operations, the measure of the crane's steadiness in relation to the tipping axis on the same side as the boom.

Impact loading: Sudden forces acting on an object due to collisions or other dynamic events.

Inclined planes: Flat surfaces placed at a shallow angle that connect levels at different heights. A ramp is an example of an inclined plane.

Momentum: A property of a moving object that is directly proportional to both its mass and speed.

Non-centered lift: Any lift attempted when the hoist line is not vertical or the hook is not directly above the load's center of gravity.

Pendulum effect: The dynamic effects of a swinging object on the support from which it is suspended.

Rebar: Shortened form of the term reinforcement bar. Used to identify individual rods, bundles, and field-fabricated meshes of rods that reinforce poured concrete structures.

Additional Resources

This module is intended as a thorough resource for task training. The following reference materials are recommended for further study.

ASME Standard B30.5, Mobile and Locomotive Cranes. Current edition. New York, NY: American Society of Mechanical Engineers.

ASME Standard B30.20, Below-the-Hook Lifting Devices. Current edition. New York, NY: American Society of Mechanical Engineers.

ASME Standard P30.1, Planning for Load Handling Activities, Current edition. New York: The American Society of Mechanical Engineers.

Bob's Rigging and Crane Handbook, Latest edition. Leawood, KS: Pellow Engineering Services.

Crane Safety on Construction Sites, 1998. Task Committee on Crane Safety on Construction Sites. Reston, VA: ASCE.

29 *CFR* 1926, Subpart CC,*Cranes and Derricks in Construction*. **www.ecfr.gov**.

Rigging Handbook, 4th Edition. 2012. Jerry A. Klinke. Stevensville, MI: ACRA Enterprises, Inc.

Willy's Signal Person & Master Rigger Handbook, First Edition. 2013. Ted Blanton Sr, Robert O'Leary, Joe Crispell, and Ted Blanton Jr. Lake Mary, FL: NorAm Productions, Inc.

Figure Credits

The Manitowoc Company, Inc., Figure 13

Link-Belt Construction Equipment Company, Table 1

© iStock.com/cosmin4000, SA02

Mammoet USA South Inc., Figure 14

Holloway Houston, Inc., Figures 26, 27

Tandemloc, Inc., Figure 28

Vestil Manufacturing, Figure 29

© iStock.com/chinaface, Figure 30

Answer	Section Reference	Objective
Section One		
1. d	1.0.0	1a
2. a	1.2.1	1b
3. b	1.3.0	1c
Section Two		
1. b	2.1.0	2a
2. a	2.2.0; Table 2	2b
3. d	2.3.1	2c
4. c	2.4.1	2d

Section Review Calculations

2.0.0 SECTION REVIEW

Question 2

$W = W_{Perp}$ = 100 lb; CF for metal against concrete = 0.60, per *Table 2*. Use these values with the formula to determine the needed force:

$F = CF \times W_{Perp}$
$F = 0.60 \times 100 \text{ lb}$
$F = 60 \text{ lb}$

The required force is **60 lb**.

Load Charts

OVERVIEW

A lift can be accomplished safely as long as the operator has a working knowledge of the forces acting on the crane at any given time during the operation. Prior to the lift, operators must know the limitations of the crane based on the load weight and the crane configuration. This important information is obtained primarily from load charts. This module explains how the information on load charts for various types of cranes is used to ensure safe lifts.

Module 21301

From *Advanced Rigger, Trainee Guide*, NCCER.
Copyright © 2018 by NCCER. Published by Pearson. All rights reserved.

Objectives

When you have completed this module, you will be able to do the following:

1. Explain how center of gravity and leverage affects crane capacity.
   a. Explain the concepts of center of gravity and leverage.
   b. Describe the content of typical load charts.
2. Describe the various crane configurations that affect load chart selection.
   a. Define and describe the four standard mobile crane configurations.
   b. Explain the significance of boom length, boom angle, and load radius.
   c. Identify the quadrants of operation and their relationship to load charts.
   d. Explain the significance of the crane's base and its configuration.
3. Describe how to apply load charts and deductions from capacity.
   a. Describe the use of jib charts.
   b. Describe how the parts of line can affect capacity.
   c. Describe how to read a range diagram.
   d. Describe how to calculate a crane's capacity from its load charts.

Performance Tasks

Under the supervision of your instructor, you should be able to do the following:

1. Identify boom angle, boom length, and load radius on a load chart.
2. Identify the requirements of the on-rubber load chart.
3. Identify the requirements of the on-outrigger load chart.
4. Properly identify load charts that are used in different configurations.
5. Identify parts of line and counterweight considerations in load chart information.
6. Calculate minimum parts of line required.

Trade Terms

Boom angle	Gross capacity	Net capacity
Boom length	Gross load	Offset
Boom-point elevation	Interpolate	Ringer crane
Configuration	Line pull	Tipping axis
Effective weight	Load operating radius	

Industry Recognized Credentials

If you are training through an NCCER-accredited sponsor, you may be eligible for credentials from NCCER's Registry. The ID number for this module is 21301. Note that this module may have been used in other NCCER curricula and may apply to other level completions. Contact NCCER's Registry at 888.622.3720 or go to **www.nccer.org** for more information.

Contents

1.0.0 GRAVITY, LEVERAGE, AND LOAD CHARTS

Objective

Explain how center of gravity and leverage affects crane capacity.

a. Explain the concepts of center of gravity and leverage.
b. Describe the contents of typical load charts.

Performance Tasks

1. Identify boom angle, boom length, and load radius on a load chart.

Trade Terms

Boom angle: In load charts, the angle between the boom relative to the horizontal plane. On telescopic-boom cranes, boom angle instruments indicate the angle between the bottom of the base section and the horizontal with the boom loaded. On lattice-boom cranes, it is the angle between the center line of the boom and a horizontal line.

Boom length: Measured from the boom foot pins to the center of the boom-point sheaves. It does not include the length of an attached jib.

Boom-point elevation: Distance from the ground to the center of the boom-point sheaves.

Configuration: A term that addresses all aspects of a crane's condition for a given lift that can include boom assembly and position, hoisting line and accessories, base arrangement, counterweights, and other factors addressed in a load chart.

Gross capacity: A crane capacity value as displayed in the crane's load chart for a specified configuration.

Gross load: The total weight of all the items lifted by the crane. The gross capacity of the crane minus any deductions for the crane configuration results in the gross load that can be lifted. Gross load is directly related to net capacity.

Load operating radius: The horizontal distance from a vertical line through the axis of crane rotation to the center of the vertical hoist line or tackle, with a load applied. Also called load radius, lifting radius, or operating radius.

Net capacity: The crane's hoisting force available after taking into account the crane's and/or load's configurations relating to a lift. Equal to load chart gross capacity minus weight deductions.

Tipping axis: In crane operations, an imaginary horizontal line around which a crane's center of gravity would rotate as it tips.

Long ago, crane operators relied on the feel of the crane as it began to tip as a capacity indicator. They would feel the crane begin to tip, and then react the best they could to keep it from overturning. Older cranes were built using components that were heavier than necessary in order to ensure safe lifts. These overbuilt cranes would tip before there was any chance of structural failure.

Today, there are much better methods available to ensure safe lifts, due to technological developments in crane construction. Cranes are now built with lighter boom components; however, this means that they are as likely to suffer structural failure as they are to tip. While relying on the feel of a crane was never a safe way to avoid tipping, an operator doing so today could experience structural failure of the crane before tipping occurs.

It is extremely important for crane operators to understand the information provided on load charts. Thousands of man-hours go into developing load charts for cranes to ensure safety. Operators must be able to distinguish between the different tables and countless numbers found on load charts, and know how to apply the information correctly when planning a lift.

This module provides information related to the use of load charts. It also describes methods used to determine a crane's gross capacity and net capacity.

1.1.0 Center of Gravity and Leverage

The principles of center of gravity and leverage play a critical role in crane operations. An understanding of these principles is essential to every lifting operation.

1.1.1 Center of Gravity

A key concept associated with lifting ability is center of gravity. The crane's center of gravity (CG) is a conceptual point where crane designers calculate the crane's weight can be located. The actual position of a crane's CG depends on the locations of the weights of all its attached components taken together. The components that contribute most to the crane's CG include the boom, carrier, upperworks, and counterweight. The load also has a CG, which riggers must know when preparing a load in order to ensure stability when it is hoisted.

The location of the crane's CG can change depending on its configuration, as shown in *Figure 1*. When the crane is in over-the-side, over-the-front, or over-the rear configurations, the center of gravity is shifted either closer to or farther away from the associated tipping axis. This, in turn, may decrease or increase the crane's leverage.

1.1.2 Leverage

Leverage is another important concept associated with crane lifting operations. The crane has leverage, and the load also exerts leverage. The load's leverage is the product of the load's weight and the shortest horizontal distance between the load's

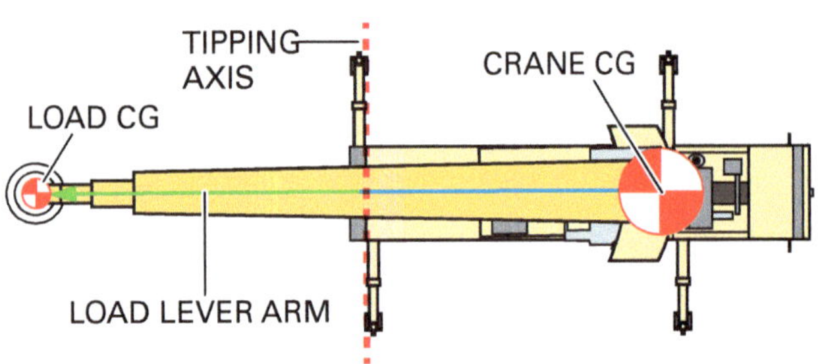

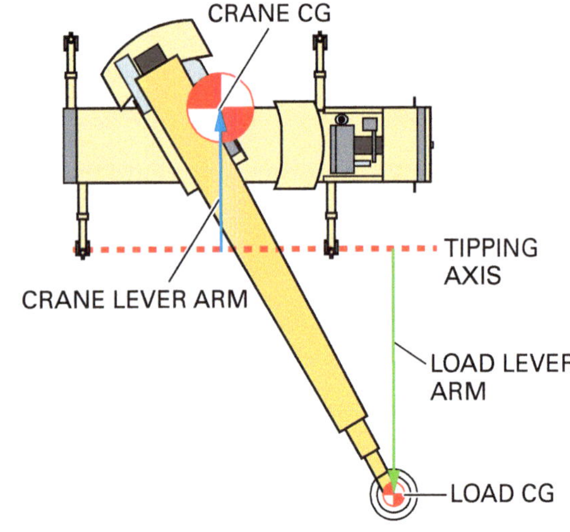

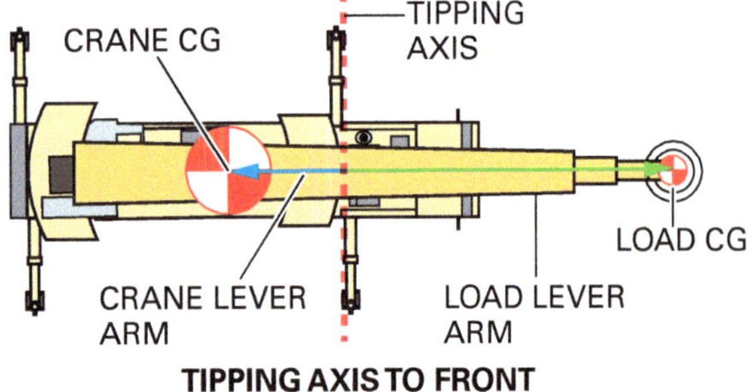

Figure 1 Typical center of gravity and leverage distances for each position of a truck-mounted crane.

CG and the tipping axis. To make a safe, stable lift, the crane's leverage has to be greater than the load's leverage (*Figure 2*). If the load's leverage is roughly equal to or greater than the crane's leverage, a tipping situation results. In addition to the configuration of the crane, booming down or telescoping out increases the load operating radius and moves the load's center of gravity further from the tipping axis. This, in turn, increases the load's leverage.

Leverage is addressed in load charts, represented by the crane's gross capacity. This is the balance of leverage remaining after subtracting the crane's weight from the leverage moment.

For a practical example of this concept, pick up a heavy book and place it directly over your head with your arms straight up. Then, keeping your arms straight, begin lowering the book from above your head until it is directly in front of you. You can quickly feel the leverage of the book. Notice how much more difficult it is to hold the book in this position due to the leverage applied by the book.

Crane and load leverages are the fundamental quantities that load charts take into account when establishing limits for safe crane operations. When you have a thorough understanding of these principles, you will see why it is essential for operators to refer to load charts before planning any lift.

1.2.0 Crane Load Charts

The crane manufacturer designs a load chart for each crane model. The charts may even be serial-number specific, meaning a unique load chart is assigned to each individual crane. This may be done if there is a slight difference between cranes due to design changes, manufacturing techniques, materials used, or permissible boom configurations. While some load charts may show the crane serial number, this isn't always the case, since some load charts apply to more than one crane. A load chart applies to only to the crane(s) specifically identified on the chart.

> **WARNING!**
>
> Load charts found online are not typically crane-specific. As a result, online charts should never be relied upon. Always locate and use crane-specific charts.

If the applicable load chart for a crane becomes lost or illegible, only the manufacturer can replace it. An individual can't legally redraw or recreate a crane's load chart for lifting purposes.

NCCER – *Advanced Rigger*

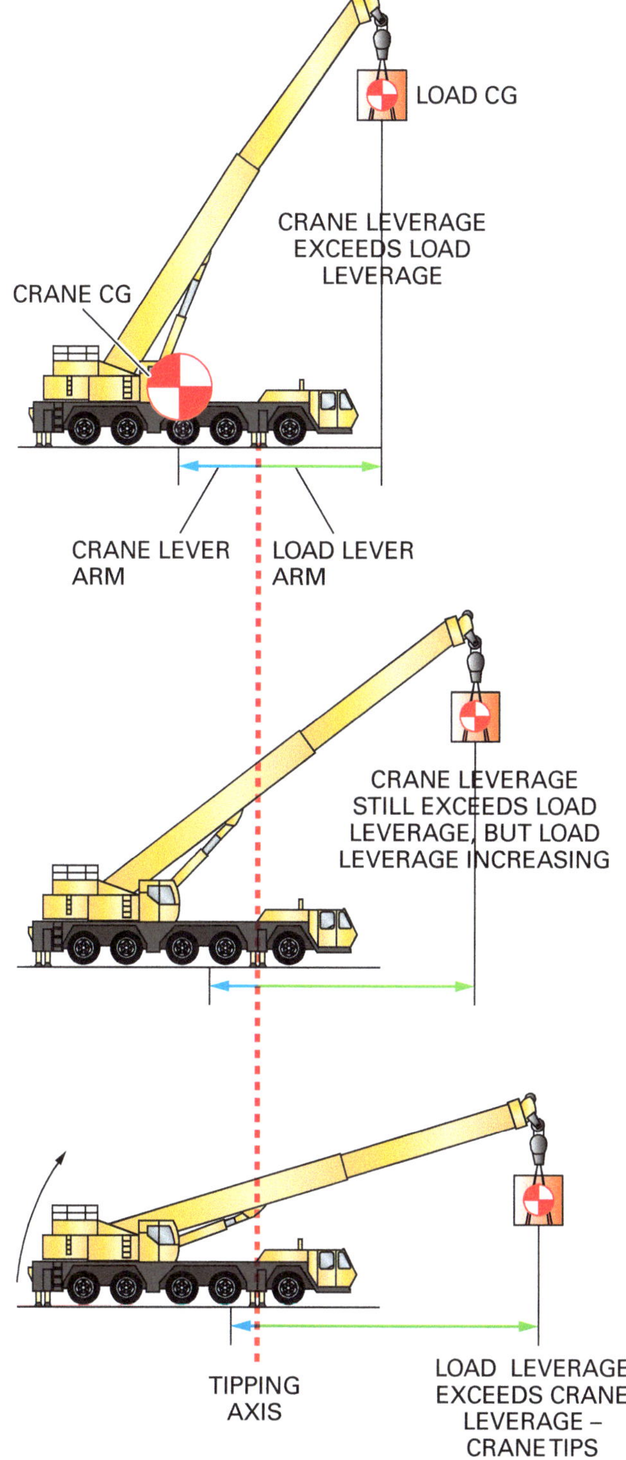

Figure 2 Relative tipping moment.

A crane cannot be operated legally without its load chart.

According to *ASME B30.5, Mobile and Locomotive Cranes*, manufacturers are required to provide specific information in load charts for a crane. Furthermore, 29 *CFR* 1926.1417(c) requires that owners post load charts in a location continuously accessible to the operator at the controls.

For situations where the operator can access only an electronic load chart, any loss of the electronic display requires that all crane operations must cease and be secured until access to the crane's load chart is restored.

Each load chart must display identifying information associating it with a specific crane(s). The complete set of load charts for a crane should generally include the following information:

- The complete range of loads for all operating configurations of the crane, including the use and non-use of any optional equipment that may be installed and that may affect the ratings
- A work area chart, which is often referred to as a *quadrant-of-operation diagram*, corresponding to the load chart in use
- Component weights, such as jibs and similar accessories
- Work areas and loads where no loads are permitted to be handled
- Specifications for lifting directions other than the most stable, if applicable
- Recommended wire rope dimension and weight limitations
- Recommended parts of line for reeving
- Load limits and precautions during boom extension and retraction, if applicable
- Tire pressures, if applicable
- Cautions, warnings, and notes relating to load limitations

> **NOTE**
> Standards no longer require load limitations based on structural and/or mechanical strength of materials, stability, hydraulics, and/or other factors. However, load charts for older cranes may include this information.

Load charts indicate the gross lifting capacity of the crane under stated conditions. This is not the weight of the load alone suspended from the crane hook. The gross load represents the total weight of all the items lifted by the crane. Net capacity, the crane's actual capacity available for a safe lift, is the gross capacity minus weight deductions related to the various crane configurations. This also represents the gross load the crane can safely lift. Refer to *Figure 3* and *Figure 4*.

The manufacturer bases load chart capacities on the maximum allowable freely suspended loads at a given boom length, boom angle, and load operating radius, as shown in *Figure 5*. Boom-point elevation relates directly to boom length and boom angle. Boom length is the length of the main boom and boom extensions, if applicable, as defined by standards. It can vary with different crane models and configurations.

Figure 3 Typical deductions for a telescopic-boom crane.

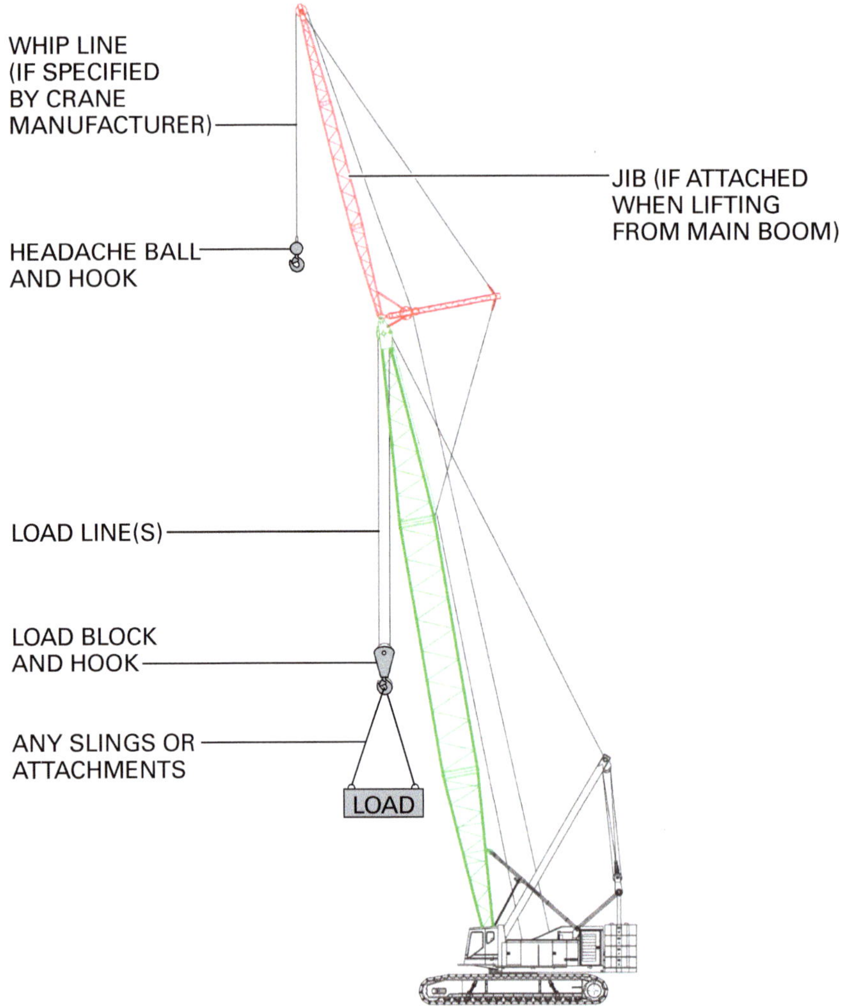

Figure 4 Typical deductions for a lattice-boom crane.

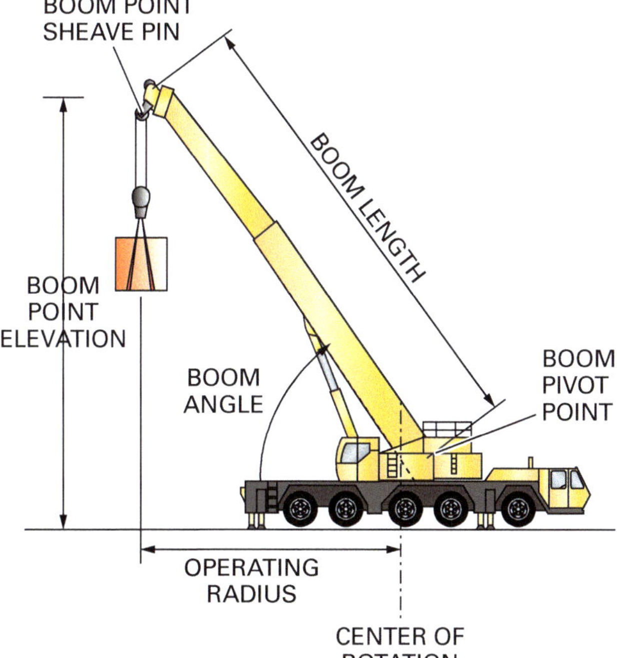

Figure 5 Boom length, boom angle, load operating radius, and boom-point elevation.

Load charts are determined for a machine standing level on a firm, supporting surface under the following ideal operating conditions:

- Equipment in original operating condition
- Equipment operated by an experienced, qualified operator
- No inclement weather during operation or with a suspended load
- Known load weight
- Correct rigging and counterweights used
- Boom length, operating radius, and boom angle known
- Load block or hook over the center of gravity of the load

Refer to the manufacturer's load charts for notes concerning allowances for conditions other than ideal. The operator must limit the load to allow for adverse conditions. Also note that load chart capacities do not exceed the allowable percentage of tipping loads. OSHA standards do not permit on-rubber or crawler capacities to exceed 75 percent of the tipping capacity as actually measured by testing. The limit is 85 percent when on outriggers.

> Crane operators working in countries other than the United States should be aware that many countries abide by ISO standards for determining crane stability. In most cases, these standards meet or exceed the 75% / 85% tipping capacity limits specified by ASME and OSHA standards. However, this is not consistent, and operators should never assume that the same standards have been met. They must carefully study the load charts provided to understand the basis for the crane's stability limits and operate accordingly.

When operators are unable to interpret a load chart correctly, they may be tempted to rely on guesswork or the sensation of tipping to determine the limits of a crane. These practices have resulted in serious accidents and fatalities. Using the onset of tipping as an indication that the crane has reached capacity is never acceptable. Structural overloading of a crane can occur before any signs of tipping are evident, resulting in a catastrophic failure. While older, overbuilt cranes are likely to tip before a structural failure occurs, there is no reason to operate them differently. The load chart must be followed regardless.

If there is no capacity listing for a particular boom length or radius, standards prohibit the operator from lifting in that configuration. Also, be aware that load charts limit lowering a boom below a minimum angle in some quadrants, because the weight of the boom and attachments may cause tipping, even without a load.

Operators must understand that load chart capacities reflect operating limits imposed by either the structural strength of the crane or its stability. The use of shading, bold lines, or asterisks in a load chart generally indicate the operating conditions allowed within these limitations. *Figure 6A* and *Figure 6B* show examples of such load charts.

Main Boom Capacities* – 110 Ft. — Open Throat Boom

Load Radius (ft)	Boom Angle (deg)	360 Rotation				Over End Blocked	
		135K + LWR (lb)	135K + 0 (lb)	83K + 0 (lb)	31K + 0 (lb)	135K + LWR (lb)	135K + 0 (lb)
18.5	82.0	346,500	346,500	329,000	283,900	346,500	346,500
19	81.7	344,400	344,400	320,000	276,000	344,400	344,400
20	81.2	341,100	341,100	303,900	246,900	341,100	341,100
25	78.6	276,100	276,100	233,200	158,900	276,100	276,100
30	75.9	229,100	226,500	171,200	116,000	229,100	229,100
35	73.2	192,600	178,500	134,500	90,500	195,200	195,200
40	70.4	158,500	146,800	110,300	73,700	169,700	169,700
50	64.8	116,200	107,400	80,100	52,900	133,700	133,000
60	58.9	90,900	83,900	62,200	40,400	109,700	102,800
70	52.6	74,100	68,300	50,200	32,100	90,700	83,100
80	45.7	62,100	57,200	41,600	26,100	75,800	69,300
90	37.9	53,100	48,800	35,200	21,700	64,700	59,100
100	28.3	46,100	42,200	30,200	18,100	56,100	51,100
110	13.8	40,400	36,900	26,100	15,200	49,200	44,700

SHADED AREA RATINGS BASED ON STRUCTURAL STRENGTH

UNSHADED AREA RATINGS BASED ON STABILITY

*Ratings are not more than 75% of tipping loads

(A) EXAMPLE FROM LINK-BELT CRANE

Boom and Jib Length	Jib Radius (ft)	5.0 Offset		15.0 Offset		25.0 Offset	
		Boom Angle	Ratings (lbs)	Boom Angle	Ratings (lbs)	Boom Angle	Ratings (lbs)
40'	45	81.0	19,380*				
(12.2M)	50	79.8	18,840*				
JIB	60	77.3	17,740*	78.9	16,740*	80.4	15,650*
&	70	74.9	16,670*	76.5	15,890*	77.9	15,050*
200'	80	72.4	15,640*	74.0	15,050*	75.4	14,440*
(61.0M)	90	69.9	13,330	71.5	13,330	72.9	13,330
BOOM	100	67.4	11,190	68.9	11,200	70.3	11,200
	110	64.7	9,480	66.3	9,480	67.6	9,490
	120	62.1	8,070	63.6	8,070	64.9	8,070
	130	59.3	6,900	60.8	6,900	62.1	6,900
	140	56.5	5,900	58.0	5,900	59.3	5,900
	150	53.6	5,040	55.1	5,040	56.3	5,040
	160	50.6	4,290	52.0	4,290	53.2	4,290
	170	47.4	3,640	48.8	3,640	49.9	3,650
	180	44.0	3,070	45.4	3,080	46.5	3,080
	190	40.4	2,560	41.8	2,560	42.8	2,570
	200	36.6	2,110	37.9	2,110	38.8	2,120

RATINGS WITH AN ASTERISK BASED ON STRUCTURAL STRENGTH

RATINGS WITHOUT AN ASTERISK BASED ON STABILITY

(B) EXAMPLE FROM TEREX CRANE

Figure 6A Structural strength and stability charts (1 of 2).

Rated Lifting Capacities in Pounds (35 Ft – 110 Ft Boom)
On Outriggers Fully Extended – 360

Radius in Feet	#0001								
	Main Boom Length in Feet								
	35	40	50	*60	70	80	90	100	110
10	100,000 (63.5)	80,400 (66.5)	74,400 (71.5)	44,600 (75.5)					
12	88,050 (60)	79,050 (63.5)	70,900 (69)	44,600 (74)	@35,600 (75.5)				
15	74,500 (54)	67,450 (59)	63,350 (65.5)	44,600 (71)	35,600 (74)	@33,000 (75.5)			
20	54,700 (43)	53,850 (50.5)	50,900 (59)	44,600 (66)	35,600 (70)	33,000 (72.5)	25,500 (75)	@23,300 (75.5)	
25	41,450 (29)	41,150 (40.5)	40,700 (52.5)	40,350 (60.5)	35,550 (65.5)	33,000 (69)	25,500 (71.5)	23,300 (74)	@18,500 (75.5)
30		32,450 (28)	32,050 (45)	31,750 (55)	30,550 (61)	28,950 (65)	25,500 (68)	23,300 (71)	18,500 (73)
35			25,950 (36.5)	25,650 (48.5)	26,500 (56.5)	24,900 (61)	23,000 (64.5)	21,200 (68)	18,500 (70.5)
40	See Note 16		21,400 (25)	21,150 (41.5)	22,000 (51.5)	21,750 (57)	20,000 (61)	18,450 (65)	18,000 (67.5)
45				17,600 (33.5)	18,500 (46)	19,100 (53)	17,600 (57.5)	16,300 (61.5)	15,750 (65)
50				14,600 (23)	15,250 (39.5)	15,700 (48)	15,650 (53.5)	14,400 (58)	13,950 (62)
55					12,650 (32.5)	13,100 (43)	13,550 (49.5)	12,850 (54.5)	12,450 (59)
60					10,500 (23)	11,000 (37.5)	11,450 (45)	11,550 (51)	11,150 (55.5)
65						9,350 (31)	9,780 (40.5)	10,200 (47)	10,050 (52.5)
70						8,870 (22)	8,370 (35)	8,780 (43)	9,090 (49)
75							7,180 (28.5)	7,590 (38.5)	7,980 (45)
80							6,120 (20)	6,560 (33)	6,950 (41)
85								5,680 (27)	6,060 (37)
90								4,910 (19)	5,280 (32)
95									4,600 (26)
100									3,990 (18.5)
Minimum boom angle (deg.) for indicated length									0
Maximum boom length (ft.) at 0 deg. boom angle (no load)									110

RATINGS ABOVE BOLD LINE IN EACH COLUMN BASED ON STRUCTURAL STRENGTH

RATINGS BELOW BOLD LINE IN EACH COLUMN BASED ON STABILITY

Note: ( ) Boom angles are in degrees.

(C) EXAMPLE FROM MANITOWOC CRANE

Figure 6B Structural strength and stability charts (2 of 2).

Additional Resources

ASME Standard B30.5, Mobile and Locomotive Cranes. Current edition. New York, NY: American Society of Mechanical Engineers.

29 *CFR* 1926.1417(c), *Cranes and Derricks in Construction – Operation*. **www.ecfr.gov**

1.0.0 Section Review

1. The leverage of a load is found by _____.

    a. multiplying the load's weight by the shortest horizontal distance between the load's CG and the tipping axis
    b. multiplying the crane's weight by the horizontal distance from the load's CG to the tipping axis
    c. dividing the load's weight by the shortest horizontal distance between the load's CG and the tipping axis
    d. dividing the crane's weight by the horizontal distance from the load's CG to the tipping axis

2. A given load chart applies to _____.

    a. all models built by a crane manufacturer
    b. all cranes of the same model built by a manufacturer
    c. all cranes that have a common design
    d. only the cranes specified by the chart

2.0.0 CRANE CONFIGURATION

Objective

Describe the various crane configurations that affect load chart selection.

a. Define and describe the four standard mobile crane configurations.
b. Explain the significance of boom length, boom angle, and load radius.
c. Identify the quadrants of operation and their relationship to load charts.
d. Explain the significance of the crane's base and its configuration.

Performance Tasks

2. Identify the requirements of the on-rubber load chart.
3. Identify the requirements of the on-outrigger load chart.

Trade Terms

Interpolate: To estimate or calculate an unknown intermediate value that falls between two known values.

Ringer crane: A large, high-capacity crane erected on a circular track (the ring, usually installed directly on the ground) that supports the entire crane base.

The capacity of a given crane depends on many factors, such as the type of crane, the type of boom and its attachments, the configuration of the crane base, and other specifics. Lift planning determines the likely boom length, boom angle, and load radius that are going to be employed during a lift. Load charts provide consideration of all of these factors, including whether the crane is lifting on rubber or on outriggers. This section addresses each of these items.

2.1.0 Types of Cranes and Hardware

There are four categories of mobile cranes for which NCCER certification is available:

- Industrial/all-purpose cranes
- Lattice-boom cranes
- Telescopic-boom cranes
- Boom trucks

These designated categories are based on the crane application, boom type, and/or vehicle type.

2.1.1 Industrial/All-Purpose Cranes

Industrial/all-purpose cranes are relatively small, lightweight, four-wheeled machines designed for industrial yard operations (*Figure 7*). They are not intended for highway operation. Manufacturers often sell this type of crane, especially the larger versions, with four-wheel crab and coordinated steering as well as the normal two-wheel front or rear capabilities. They are operated from the driver's seat, and have a telescoping-boom crane mounted on the chassis. They are suitable for on-rubber operations, and some are designed for pick-and-carry work, as well as lifts on outriggers. The boom upperworks can usually rotate a full 360 degrees.

Many industrial/all-purpose cranes include a flat, load-carrying deck in their design. Note that the term *pick-and-carry* refers to moving a load on the hook, not on the deck. Those that can only carry a load on the deck and not on the hook are often referred to as *carry-deck cranes*.

Figure 7 Industrial/all-purpose crane.

2.1.2 Lattice-Boom Cranes

Lattice-boom cranes can be truck-mounted or crawler-mounted (*Figure 8*). Because lattice booms are strong but relatively lightweight, they can be constructed or extended to lengths up to several hundred feet—longer than equal-weight telescopic booms. Lattice-boom cranes also vary in the types of load control systems used and the kind of work the crane does. Most modern models and those used for straight lifting work use hydraulic motors and controls. Older model cranes and those used for large dynamic loads, such as pile-pulling and driving, use friction clutches and brakes. These are known as *friction cranes*. Site preparation and erection of lattice-boom cranes may take time—from a day for smaller rigs, to weeks for larger, heavy-lift units with their various attachments.

2.1.3 Telescopic-Boom Cranes

Telescopic-boom cranes are categorized for NCCER certifications by carrier type—rough/all-terrain (RT/AT) cranes (*Figure 9*), tire-mounted truck cranes, and crawler cranes. It is common to further categorize RT/AT types by whether the control station rotates with the boom or is stationary. For larger cranes, the operator's cab usually rotates with the upperworks supporting the boom. Hydraulic components operate and extend all telescopic booms. Because of their construction, they are generally shorter for a given boom weight, compared to lattice-boom cranes.

2.1.4 Boom Trucks

Boom trucks (*Figure 10*) consist of a telescopic- or articulating-boom upperworks placed on a tire-mounted carrier designed and engineered for highway travel. Certifications in this category designate fixed or rotating control stations, and whether the boom is of the telescoping or articulating type. Boom trucks are used in many applications where highly mobile units with limited capacity are needed. Models with special insulation systems are regularly used to service power distribution systems.

2.1.5 Crane Configuration Considerations

Manufacturers base load charts of their mobile cranes on the configuration of the crane at the time of the lift. Some examples of configurations are:

- Jibs and boom extensions not installed
- Jib and/or boom extensions installed but load lifted from main boom

Figure 8 Truck-mounted and crawler-mounted lattice-boom cranes

- Jibs and/or boom extensions installed and load lifted from jib or boom extension

The principal factors influencing the machine's capacity and correct interpretation of the chart include these considerations:

- Boom length, boom angle, and load radius

NCCER – *Advanced Rigger*

Figure 9 Telescopic-boom crane with rotating control station.

Figure 10 Boom truck.

- Type of boom
  - Telescopic
  - Lattice
- Operational quadrant in which the superstructure is working
- Configuration of crane base
  - On rubber
  - On outriggers (stabilizers)
  - On crawlers
- Telescopic boom modes (A or B)
- Effective/stowed weight of jib or boom extension (or other attachments)
- Ring or tower attachment
- Type, size, and weight of load block and headache ball
- Weight of all required rigging
- Weight of rope per manufacturer's requirement
- Counterweight configuration
- Parts of line

2.2.0 Boom Length, Boom Angle, and Load Radius

The load radius for telescopic- and lattice-boom cranes can be found in various locations in a set of load charts. The location of the boom angle information can also vary. Understanding where these resources are on a load chart is key to reading and properly interpreting a load chart. Operators must demonstrate their ability to read, comprehend, and write in the language of the crane manufacturer's operation and maintenance documentation. They must also exhibit arithmetic skills and the ability to use the manufacturer's load charts.

Most telescopic-boom crane load charts (*Figure 11*) start with the shortest boom length at the upper left region and read to the right and down for progressively longer booms. Start by locating the number that corresponds to the length (in feet) of the boom in the Boom Length heading. Then, locate the operating radius of the crane in the left- or right-hand column. The capacity where the operating radius row and an operating quadrant under the Boom Length heading intersect is the gross capacity with the indicated counterweight configuration.

Lattice-boom crawler crane load charts may show their information differently, depending on the manufacturer. In the example chart shown in *Figure 12*, the boom length is in the chart heading, and the load radius is in the left-hand column.

The operator must first know what boom length the crane is operating at, determine the required radius for the lift, and then read across to determine the gross capacity, depending on the counterweight configuration. Some manufacturers may also include the boom elevation point in their charts for reference, especially if they require consideration of the weight of suspended hoist cable in the load calculations.

Note that the load charts shown in this module are examples from certain manufacturers. The standards do not specify a mandatory, universal format for presenting this information. Load chart layouts are as diverse as the number of manufacturers that build cranes. They can even vary greatly between different crane models from the same manufacturer. Some have illustrations and counterweight requirements at the top; some express their load radius or boom lengths in feet and/or meters (*Figure 13*, *Figure 14*, and *Figure 15*), and some list their capacities in pounds, kilopound-force (kip), or kilograms (*Figure 13* and *Figure 15*). This variety presents a challenge to operators; they must pay close attention to load charts and the operating notes when switching between cranes to avoid making mistakes on load chart calculations and interpretations.

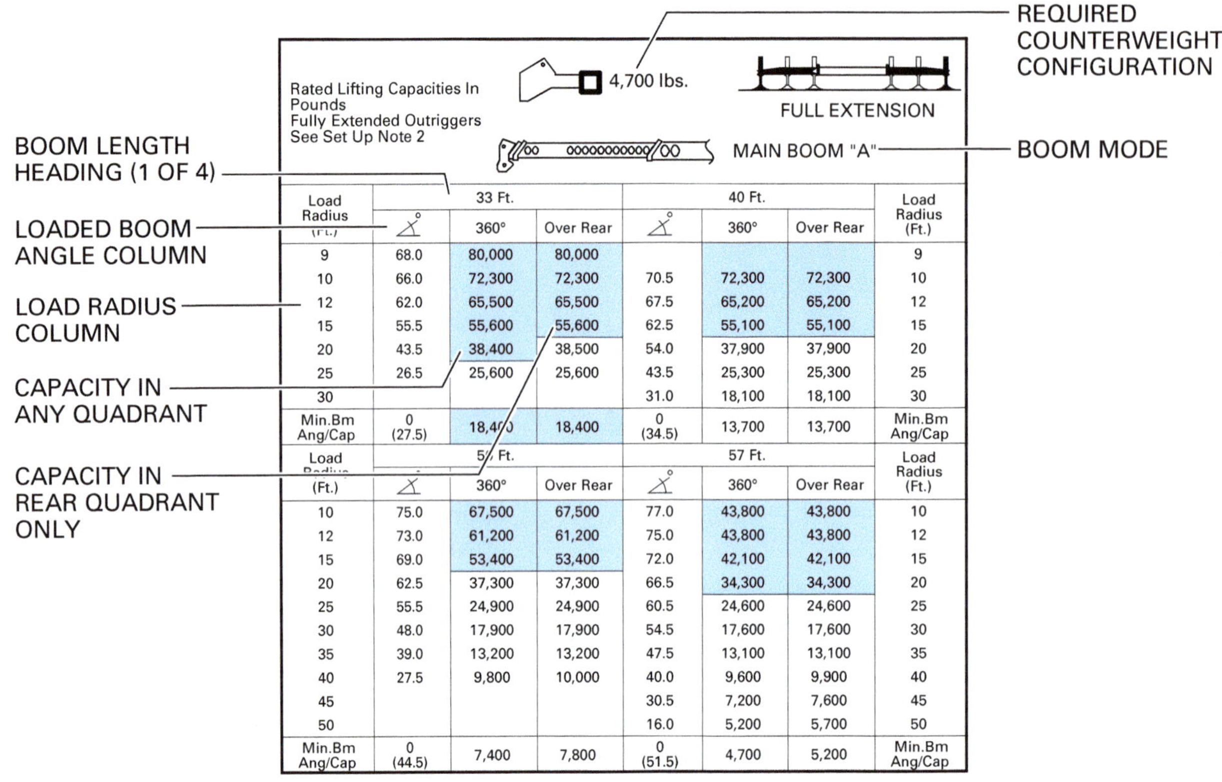

Load Radius (Ft.)	∠°	33 Ft. 360°	Over Rear	∠°	40 Ft. 360°	Over Rear	Load Radius (Ft.)
9	68.0	80,000	80,000				9
10	66.0	72,300	72,300	70.5	72,300	72,300	10
12	62.0	65,500	65,500	67.5	65,200	65,200	12
15	55.5	55,600	55,600	62.5	55,100	55,100	15
20	43.5	38,400	38,500	54.0	37,900	37,900	20
25	26.5	25,600	25,600	43.5	25,300	25,300	25
30				31.0	18,100	18,100	30
Min.Bm Ang/Cap	0 (27.5)	18,400	18,400	0 (34.5)	13,700	13,700	Min.Bm Ang/Cap

Load Radius (Ft.)	∠°	50 Ft. 360°	Over Rear	∠°	57 Ft. 360°	Over Rear	Load Radius (Ft.)
10	75.0	67,500	67,500	77.0	43,800	43,800	10
12	73.0	61,200	61,200	75.0	43,800	43,800	12
15	69.0	53,400	53,400	72.0	42,100	42,100	15
20	62.5	37,300	37,300	66.5	34,300	34,300	20
25	55.5	24,900	24,900	60.5	24,600	24,600	25
30	48.0	17,900	17,900	54.5	17,600	17,600	30
35	39.0	13,200	13,200	47.5	13,100	13,100	35
40	27.5	9,800	10,000	40.0	9,600	9,900	40
45				30.5	7,200	7,600	45
50				16.0	5,200	5,700	50
Min.Bm Ang/Cap	0 (44.5)	7,400	7,800	0 (51.5)	4,700	5,200	Min.Bm Ang/Cap

Note: Refer to page 5 for (Capacity Deductions for Auxiliary Load Handling Equipment).

∠° Loaded Boom Angle in Degrees

( ) Reference radius for minimum boom angle capacities (shown in parenthesis) are in feet.

Figure 11 Telescopic-boom crane load chart.

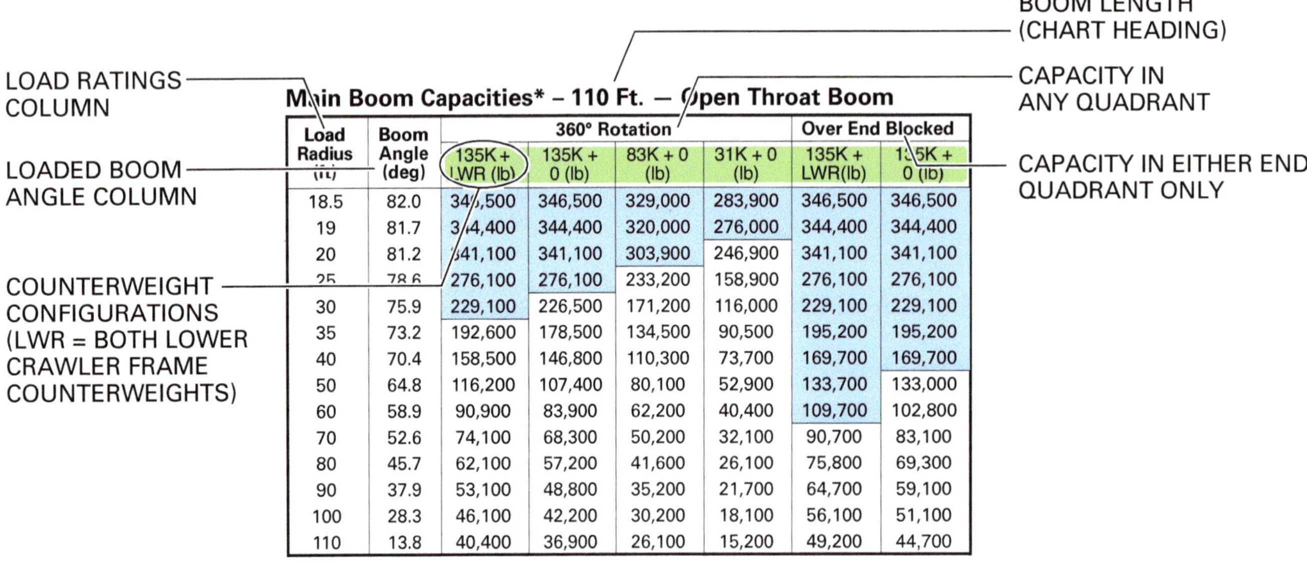

Main Boom Capacities* – 110 Ft. — Open Throat Boom

Load Radius (ft)	Boom Angle (deg)	360° Rotation 135K + LWR (lb)	135K + 0 (lb)	83K + 0 (lb)	31K + 0 (lb)	Over End Blocked 135K + LWR(lb)	135K + 0 (lb)
18.5	82.0	346,500	346,500	329,000	283,900	346,500	346,500
19	81.7	344,400	344,400	320,000	276,000	344,400	344,400
20	81.2	341,100	341,100	303,900	246,900	341,100	341,100
25	78.6	276,100	276,100	233,200	158,900	276,100	276,100
30	75.9	229,100	226,500	171,200	116,000	229,100	229,100
35	73.2	192,600	178,500	134,500	90,500	195,200	195,200
40	70.4	158,500	146,800	110,300	73,700	169,700	169,700
50	64.8	116,200	107,400	80,100	52,900	133,700	133,000
60	58.9	90,900	83,900	62,200	40,400	109,700	102,800
70	52.6	74,100	68,300	50,200	32,100	90,700	83,100
80	45.7	62,100	57,200	41,600	26,100	75,800	69,300
90	37.9	53,100	48,800	35,200	21,700	64,700	59,100
100	28.3	46,100	42,200	30,200	18,100	56,100	51,100
110	13.8	40,400	36,900	26,100	15,200	49,200	44,700

***Ratings are not more than 75% of tipping loads**

Figure 12 Lattice-boom crawler crane load chart.

Outriggers Fully Extended
25' - 43' (7.62m - 13.11m) Offset Telescoping Fly + 70.25' (21.41m) Main Boom with Manual Retracted

LOAD RADIUS (ft/m)	25' (7.62m) Fly						43' (13.11m) Fly						LOAD RADIUS (ft/m)
	2 OFFSET		15 OFFSET		30 OFFSET		2 OFFSET		15 OFFSET		30 OFFSET		
	LOADED BOOM ANGLE (deg.)	LOAD (lbs/kgs)	LOADED BOOM ANGLE (deg.)	LOAD (lbs/kgs)	LOADED BOOM ANGLE (deg.)	LOAD (lbs/kgs)	LOADED BOOM ANGLE (deg.)	LOAD (lbs/kgs)	LOADED BOOM ANGLE (deg.)	LOAD (lbs/kgs)	LOADED BOOM ANGLE (deg.)	LOAD (lbs/kgs)	
20 / 6.10	78.5	12,700 / 5,761											20 / 6.10
25 / 7.62	75.5	11,500 / 5,216	79.0	8,800 / 3,992			77.5	4,900 / 2,223					25 / 7.62
30 / 9.14	72.5	10,600 / 4,808	76.0	8,200 / 3,720	79.0	6,100 / 2,767	75.5	4,700 / 2,132					30 / 9.14
35 / 10.67	69.5	9,600 / 4,355	73.0	7,500 / 3,402	76.0	5,700 / 2,586	73.0	4,500 / 2,041	78.0	4,000 / 1,614			35 / 10.67
40 / 12.19	66.0	8,700 / 3,946	69.5	6,900 / 3,130	72.5	5,400 / 2,449	70.0	4,300 / 1,950	75.5	3,900 / 1,769			40 / 12.19
45 / 13.72	63.0	8,000 / 3,629	66.0	6,400 / 2,903	69.0	5,100 / 2,313	67.5	4,100 / 1,860	72.5	3,700 / 1,678	77.5	3,000 / 1,361	45 / 13.72
50 / 15.24	59.5	7,400 / 3,357	62.5	5,900 / 2,676	65.5	4,800 / 2,177	65.0	3,900 / 1,769	70.0	3,500 / 1,588	74.5	2,900 / 1,315	50 / 15.24
55 / 16.76	56.0	6,500 / 2,948	59.0	5,300 / 2,404	62.0	4,500 / 2,041	62.0	3,800 / 1,724	67.0	3,300 / 1,497	71.5	2,700 / 1,225	55 / 16.76
60 / 18.29	52.0	5,400 / 2,449	55.0	4,900 / 2,223	58.0	4,200 / 1,905	59.5	3,700 / 1,678	64.0	3,100 / 1,406	68.5	2,600 / 1,179	60 / 18.29
65 / 19.81	48.0	4,500 / 2,041	51.0	4,500 / 2,041	53.5	3,900 / 1,769	56.5	3,500 / 1,588	61.0	2,900 / 1,315	65.5	2,500 / 1,134	65 / 19.81
70 / 21.34	43.5	3,700 / 1,678	46.5	3,900 / 1,769	49.0	3,600 / 1,633	53.0	3,200 / 1,452	58.0	2,800 / 1,270	62.5	2,400 / 1,089	70 / 21.34
75 / 22.86	38.5	3,000 / 1,361	41.5	3,200 / 1,452	44.0	3,400 / 1,542	50.5	3,000 / 1,361	54.5	2,600 / 1,179	59.0	2,300 / 1,043	75 / 22.86
80 / 24.38	33.0	2,500 / 1,134	36.0	2,600 / 1,179	37.5	2,700 / 1,225	46.5	2,800 / 1,270	51.0	2,500 / 1,134	55.0	2,200 / 998	80 / 24.38
85 / 25.91	26.5	2,000 / 907	29.0	2,100 / 953	30.0	2,100 / 953	43.0	2,700 / 1,225	47.5	2,400 / 1,089	51.0	2,100 / 953	85 / 25.91
90 / 27.43	17.5	1,600 / 726	19.5	1,600 / 726	17.0	1,600 / 726	38.5	2,300 / 1,043	43.0	2,300 / 1,043	46.5	2,000 / 907	90 / 27.43
95 / 28.96							34.0	2,000 / 907	38.5	2,200 / 998	41.5	2,000 / 907	95 / 28.96
100 / 30.48							28.5	1,600 / 726	33.0	1,800 / 816	35.5	1,900 / 862	100 / 30.48
105 / 32.00									25.5	1,400 / 635	26.0	1,400 / 635	105 / 32.00

Warning: Do not lower 25' offset telescoping fly in working position below 12 unless main boom is 68' or less, since loss of stability will occur causing a tipping condition.

Warning: Do not lower 43' offset telescoping fly in working position below 20 unless main boom is 65' or less, since loss of stability will occur causing a tipping condition.

Figure 13 Example load chart 1.

NOTE

If you are not using a load moment indicator (LMI) but operating from load charts alone, enter the chart at the next highest load radius or next lowest boom angle; do not interpolate chart data. LMIs may mathematically interpolate the gross capacity in the charts as programmed by the manufacturer, but interpolation must not be done by the operator. Always use the lower value on a chart when a value falls between chart values.

If a load radius, boom length, or boom angle falls between values listed on the load chart, the operator cannot interpolate the crane's gross capacity. When either boom length or radius, or both, are between the values listed, the operator shall use the smallest load value shown at either the next larger radius or next longer or shorter boom length.

NOTE

Some telescoping cranes are equipped with multiple boom-extension modes. The operator selects the mode with a switch. Depending on the manufacturer, these modes may be identified by letters, numbers, or Roman numerals (*Figure 16*). The first extension mode generally involves all sections of the boom extended equally to the boom's maximum length. Other modes may either extend all sections at smaller percentages equally, or involve extending one or more of the sections by certain percentages. By delaying the

m	24 m	30 m	36 m	42 m	48 m	54 m	60 m	66 m	72 m	78 m	84 m	90 m	96 m	m
	t	t	t	t	t	t	t	t	t	t	t	t	t	
5,5	625,0	-	-	-	-	-	-	-	-	-	-	-	-	5,5
6	574,0	577,0	-	-	-	-	-	-	-	-	-	-	-	6
7	512,0	463,0	423,0	388,0	-	-	-	-	-	-	-	-	-	7
8	420,0	386,0	357,0	331,0	308,0	288,0	-	-	-	-	-	-	-	8
9	356,0	331,0	308,0	288,0	270,0	254,0	239,0	-	-	-	-	-	-	9
10	308,0	288,0	270,0	254,0	239,0	226,0	213,0	202,0	192,0	-	-	-	-	10
11	267,5	257,5	243,0	229,5	216,5	205,0	194,0	184,5	175,5	166,0	158,0	-	-	11
12	227,0	227,0	216,0	205,0	194,0	184,0	175,0	167,0	159,0	152,0	144,0	139,0	132,0	12
14	177,0	176,0	175,0	170,0	162,0	155,0	147,0	141,0	134,0	129,0	123,0	118,0	112,0	14
16	143,0	142,0	142,0	140,0	138,0	132,0	126,0	121,0	115,0	111,0	106,0	102,0	97,5	16
18	120,0	119,0	118,0	117,0	116,0	115,0	109,0	105,0	100,0	97,0	92,0	89,5	85,0	18
20	103,0	101,0	100,0	99,5	98,5	97,5	96,5	92,5	88,0	85,0	81,0	78,5	74,5	20
22	89,5	88,5	87,0	85,5	85,0	84,0	83,0	82,0	78,0	75,5	71,5	69,5	65,5	22
23	84,0	83,0	81,7	80,2	79,5	78,5	77,5	76,7	73,7	71,2	67,5	65,7	62,0	23
24	-	77,5	76,5	75,0	74,0	73,0	72,0	71,5	69,5	67,0	63,5	62,0	58,5	24
26	-	69,0	67,5	66,0	65,0	64,0	63,0	62,5	61,0	60,0	56,5	55,0	52,0	26
28	-	62,0	60,5	59,0	58,0	56,5	55,5	55,0	53,5	53,0	50,5	49,6	46,4	28
30	-	-	54,5	53,0	51,5	50,5	49,4	48,9	47,6	47,1	45,7	44,5	41,5	30
33	-	-	47,3	45,8	44,4	43,2	41,9	41,2	39,5	38,9	37,6	37,2	35,2	33
34	-	-	-	43,4	42,1	40,8	39,4	38,7	37,2	36,5	35,0	34,8	33,1	34
38	-	-	-	36,3	34,6	33,0	31,4	30,7	29,1	28,4	26,9	26,6	25,3	38
42	-	-	-	-	28,7	26,9	25,2	24,4	22,8	22,0	20,5	20,2	18,8	42
44	-	-	-	-	26,2	24,5	22,6	21,8	20,2	19,4	17,9	17,6	16,2	44
46	-	-	-	-	-	22,1	20,3	19,4	17,7	16,9	15,4	15,1	13,7	46
49	-	-	-	-	-	19,2	17,3	16,4	14,6	13,8	12,2	11,9	10,4	49
50	-	-	-	-	-	-	16,3	15,4	13,6	12,8	11,2	10,9	9,4	50
54	-	-	-	-	-	-	13,1	12,0	10,1	9,3	7,7	7,3	5,9	54
55	-	-	-	-	-	-	-	11,3	9,4	8,5	6,9	6,5	5,1	55
57	-	-	-	-	-	-	-	9,9	8,0	7,1	5,5	5,1	-	57
58	-	-	-	-	-	-	-	9,3	7,3	6,4	-	-	-	58
59	-	-	-	-	-	-	-	8,7	6,7	5,8	-	-	-	59
60	-	-	-	-	-	-	-	-	6,1	5,2	-	-	-	60
61	-	-	-	-	-	-	-	-	5,5	-	-	-	-	61

125 t   165 t

NOTE: Data published herein is intended as a guide only and shall not be construed to warrant applicability for lifting purposes. Crane operation is subject to the computer charts and operation manual both supplied with the crane.

Figure 14 Example load chart 2.

extension of the outermost sections, the boom has a greater capacity at shorter boom lengths than it would if all sections are extended the same percentage. The different configurations will result in different lifting capacities, which, in turn, necessitate different load charts for each mode.

2.3.0 Quadrants of Operation

Crane operational quadrants are horizontal sectors surrounding the crane, defined by the tipping axes for a given configuration and the direction of the crane boom. The quadrants represent variations of crane stability for positions of the boom for which the load chart must account. As a boom swings to various quadrants with no change in boom angle, the distance between the crane's or load's CGs and the nearest tipping axis changes dramatically. This is because mobile cranes, especially wheeled ones, are typically longer than they are wide. This rectangular shape tends to result in more stability of the crane from front-to-rear than from side-to-side, though the locations of tires, outriggers, and stabilizers affect the shape of operational quadrants. It is not unusual for a crane to have different capacities from one quadrant to another.

The operational quadrants are over-the-front, over-the-rear, and over-the-side (right or left) (*Figure 17*). Note that manufacturers may differ on the range of the defined arc, although 6 degrees in each direction is common.

ON OUTRIGGERS FULLY EXTENDED 23' 7-1/2" (7.2 M) SPREAD
360° ROTATION

A / B	37.7' (11.5 m)		51' (15.56 m)		64.4' (19.62 m)				91' (27.75 m)				117.7' (35.87 m)				131' (39.93 m)		144.4' (44.0 m)	
B	C		C		C		C		C		C		C		C		C		C	
10	68	160,000	75	103,600	78	88,100	78	44,000												
12	65	125,000	72	103,600	76	88,100	76	44,000												
15	60	108,000	69	103,600	73	88,100	73	44,000	79	44,000	79	30,800								
20	50	78,400	63	77,800	69	71,900	69	44,000	76	44,000	76	30,800	79	30,800	79	17,600				
25	38	59,400	56	59,000	64	56,100	64	44,000	73	44,000	73	30,800	77	30,800	77	17,600	79	17,600	80	17,600
30	21	45,900	48	44,600	59	42,600	59	44,000	70	39,000	70	26,700	75	30,800	75	17,600	77	17,600	78	17,600
35			39	33,800	53	33,000	53	39,900	66	34,000	66	23,200	72	28,200	72	17,600	75	17,600	76	17,600
40			28	26,300	47	25,500	47	32,300	63	28,700	63	20,400	70	24,700	70	17,600	73	17,600	74	17,600
45			5	20,900	40	20,000	40	26,400	59	23,600	59	18,200	67	21,800	67	16,400	71	17,600	73	17,600
50					32	15,900	32	21,900	55	19,300	56	16,400	65	19,500	64	14,700	68	16,200	71	17,100
60									46	13,200	47	14,500	59	14,800	59	11,900	63	13,300	66	13,900
70									36	9,000	37	11,400	53	10,700	53	9,900	58	11,100	62	10,900
80									22	6,100	24	9,500	46	7,600	46	8,400	53	9,000	57	8,200
90													38	5,300	38	7,200	46	6,900	51	6,100
100													27	3,500	28	5,900	39	5,100	45	4,300
110													13	2,100	12	4,600	31	3,600	39	2,900
120																	19	2,600	32	1,800
																	18°		32°	

...ions (%)

| | | | I | | II | | II | | |

Figure 15 Example load chart 3.

Many load charts cover 360 degrees of operation. However, crane manufacturers sometimes provide separate load charts for certain quadrants due to variances in capacity. For example, notice the capacity differences in the load charts in *Figure 18* for a telescopic-boom crane.

With a boom length of 36 feet, the gross capacity remains the same in both configurations at an operating radius of 10 feet. However, at all other boom lengths and operating radiuses, the capacity for 360 degrees is less than that for over-the-front operation. This is because the crane leverage is much larger in this quadrant alone than can be allowed for in 360-degree operation.

2.4.0 Configuration of the Crane Base

The crane can be set up on rubber, outriggers, or stabilizers, and with crawlers extended or retracted. Each configuration affects capacity, requiring a different load chart for each base configuration.

2.4.1 On Rubber

Figure 19 shows on-rubber load charts for various operations. Load charts assume specified tire pressures for determining gross capacities. Manufacturers must identify the required tire pressures in notes on the load charts or in a separate table.

There are several requirements for on-rubber lifting. Capacities for on-rubber lifting printed on the load chart do not exceed 75 percent of tipping loads as determined by test. This percentage is strictly an engineering limit established by American safety standards, used by the manufacturer in creating load-chart capacities. It is important to emphasize that this does not mean that the operator has a 25-percent cushion to work with. The operator must not exceed published load chart capacities under any circumstances.

2.4.2 On Outriggers

Load chart capacities for on-outrigger lifting do not exceed 85 percent of tipping loads. As with on-rubber lifting, this is an engineering safety limit. It does not mean the operator has a 15-percent cushion to work within. Depending on the design

Boom Extend Modes

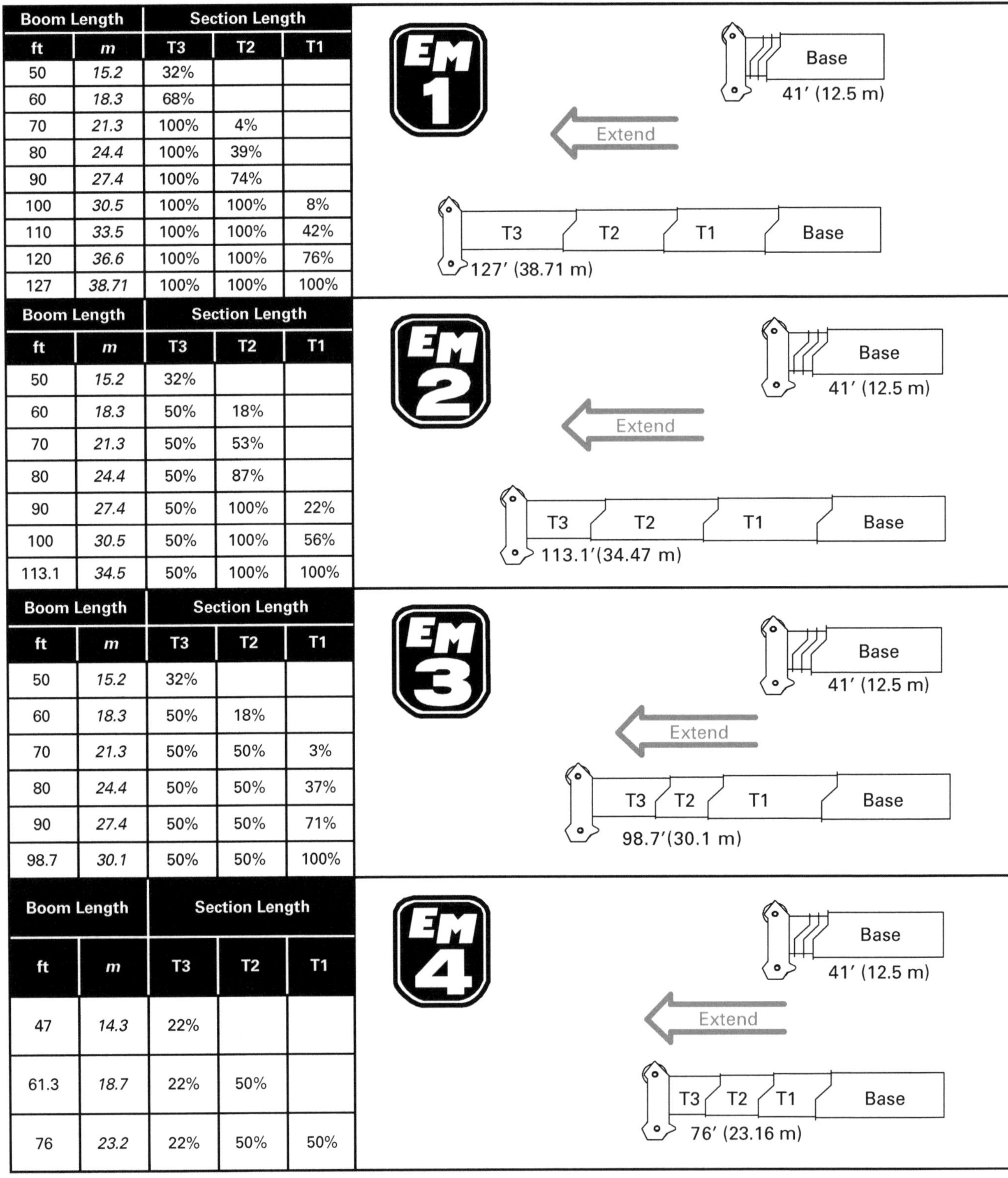

EM 1

Boom Length		Section Length		
ft	*m*	T3	T2	T1
50	*15.2*	32%		
60	*18.3*	68%		
70	*21.3*	100%	4%	
80	*24.4*	100%	39%	
90	*27.4*	100%	74%	
100	*30.5*	100%	100%	8%
110	*33.5*	100%	100%	42%
120	*36.6*	100%	100%	76%
127	*38.71*	100%	100%	100%

EM 2

Boom Length		Section Length		
ft	*m*	T3	T2	T1
50	*15.2*	32%		
60	*18.3*	50%	18%	
70	*21.3*	50%	53%	
80	*24.4*	50%	87%	
90	*27.4*	50%	100%	22%
100	*30.5*	50%	100%	56%
113.1	*34.5*	50%	100%	100%

EM 3

Boom Length		Section Length		
ft	*m*	T3	T2	T1
50	*15.2*	32%		
60	*18.3*	50%	18%	
70	*21.3*	50%	50%	3%
80	*24.4*	50%	50%	37%
90	*27.4*	50%	50%	71%
98.7	*30.1*	50%	50%	100%

EM 4

Boom Length		Section Length		
ft	*m*	T3	T2	T1
47	*14.3*	22%		
61.3	*18.7*	22%	50%	
76	*23.2*	22%	50%	50%

Figure 16 Telescopic-boom extension mode chart.

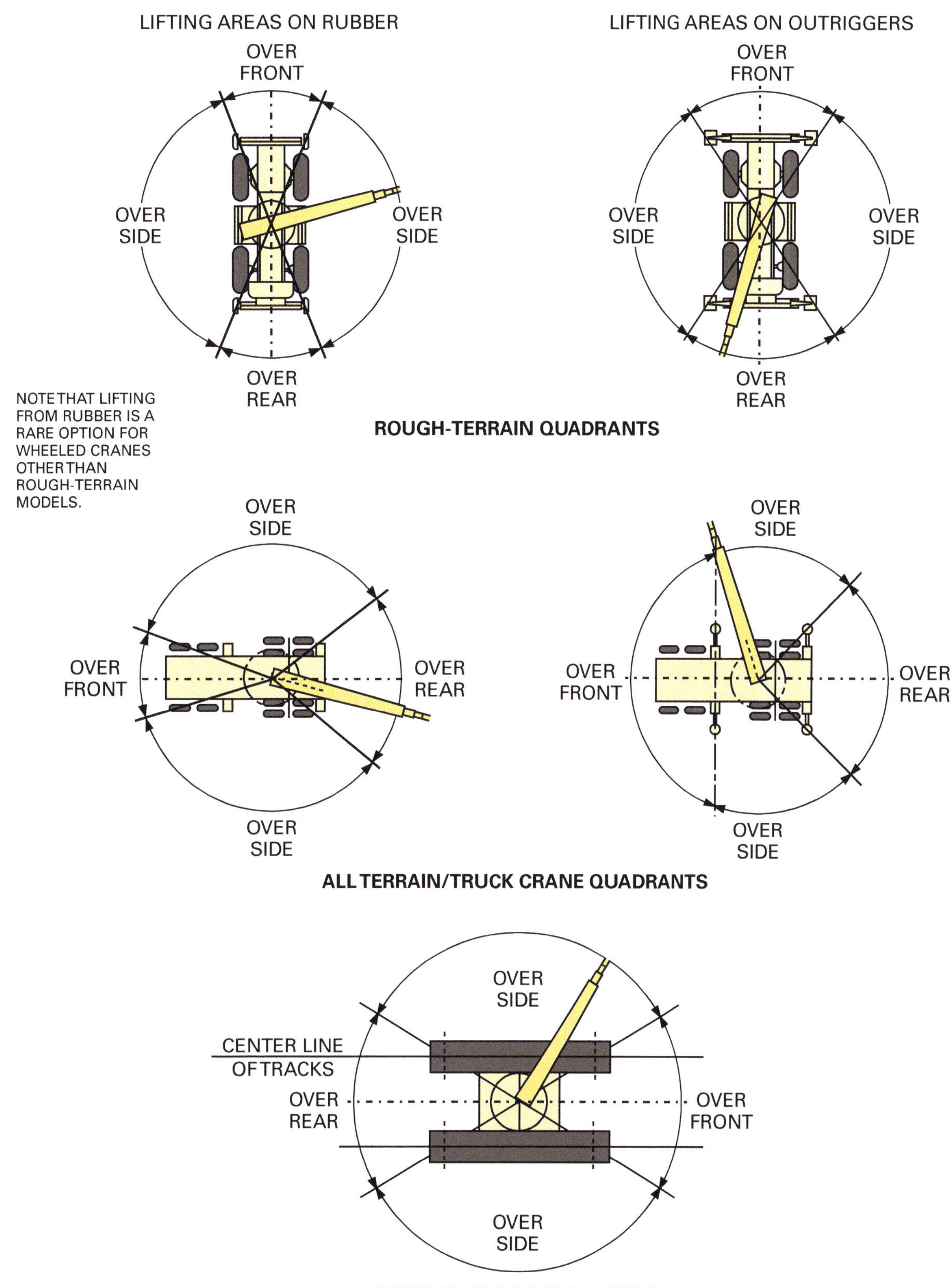

Figure 17 Quadrant definition drawings for rough-terrain, truck, and crawler cranes.

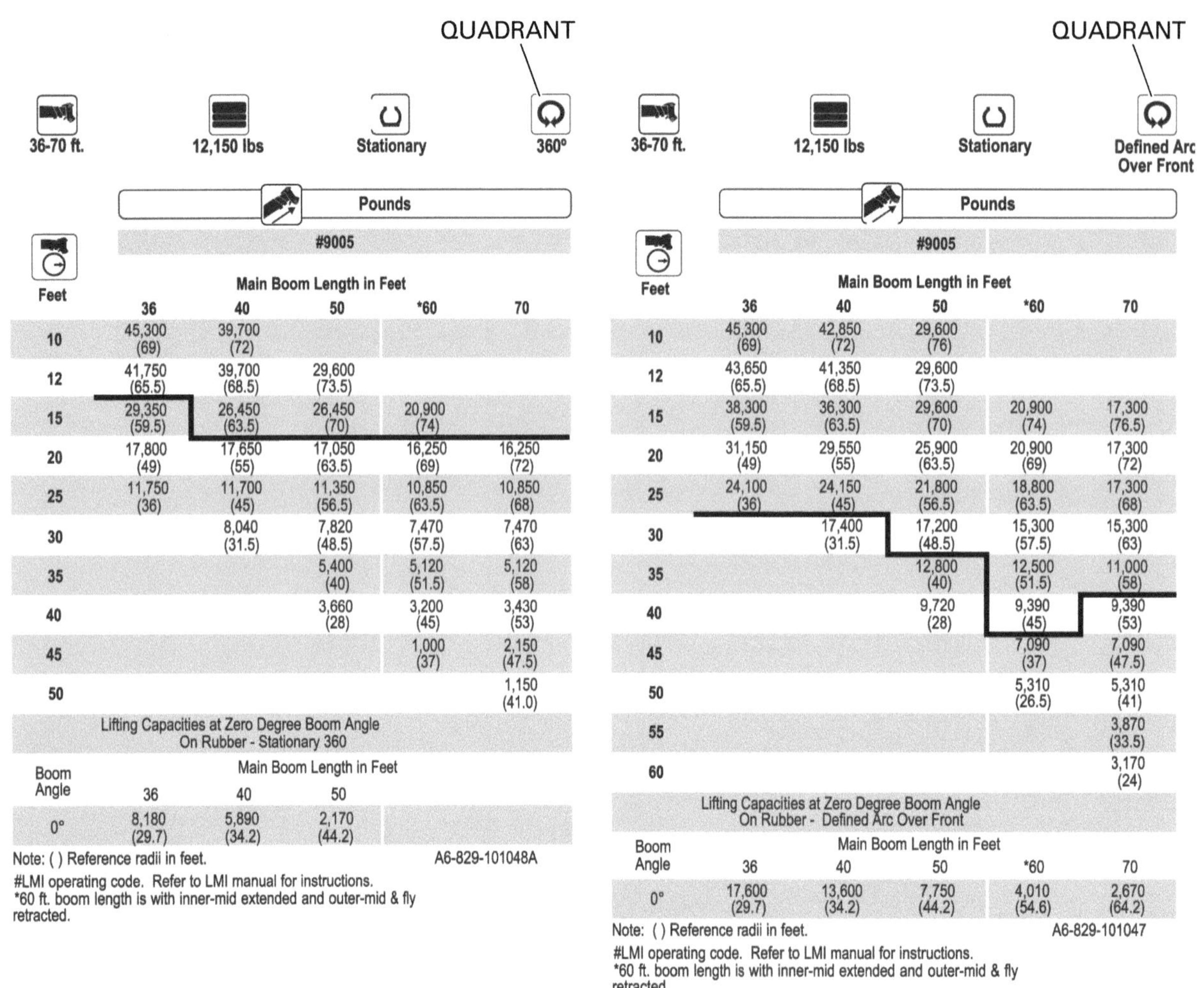

Left chart — Pounds — #9005

Feet	Main Boom Length in Feet				
	36	40	50	*60	70
10	45,300 (69)	39,700 (72)			
12	41,750 (65.5)	39,700 (68.5)	29,600 (73.5)		
15	29,350 (59.5)	26,450 (63.5)	26,450 (70)	20,900 (74)	
20	17,800 (49)	17,650 (55)	17,050 (63.5)	16,250 (69)	16,250 (72)
25	11,750 (36)	11,700 (45)	11,350 (56.5)	10,850 (63.5)	10,850 (68)
30		8,040 (31.5)	7,820 (48.5)	7,470 (57.5)	7,470 (63)
35			5,400 (40)	5,120 (51.5)	5,120 (58)
40			3,660 (28)	3,200 (45)	3,430 (53)
45				1,000 (37)	2,150 (47.5)
50					1,150 (41.0)

Lifting Capacities at Zero Degree Boom Angle
On Rubber - Stationary 360

Boom Angle	Main Boom Length in Feet		
	36	40	50
0°	8,180 (29.7)	5,890 (34.2)	2,170 (44.2)

Note: ( ) Reference radii in feet. A6-829-101048A

#LMI operating code. Refer to LMI manual for instructions.
*60 ft. boom length is with inner-mid extended and outer-mid & fly retracted.

Right chart — Pounds — #9005

Feet	Main Boom Length in Feet				
	36	40	50	*60	70
10	45,300 (69)	42,850 (72)	29,600 (76)		
12	43,650 (65.5)	41,350 (68.5)	29,600 (73.5)		
15	38,300 (59.5)	36,300 (63.5)	29,600 (70)	20,900 (74)	17,300 (76.5)
20	31,150 (49)	29,550 (55)	25,900 (63.5)	20,900 (69)	17,300 (72)
25	24,100 (36)	24,150 (45)	21,800 (56.5)	18,800 (63.5)	17,300 (68)
30		17,400 (31.5)	17,200 (48.5)	15,300 (57.5)	15,300 (63)
35			12,800 (40)	12,500 (51.5)	11,000 (58)
40			9,720 (28)	9,390 (45)	9,390 (53)
45				7,090 (37)	7,090 (47.5)
50				5,310 (26.5)	5,310 (41)
55					3,870 (33.5)
60					3,170 (24)

Lifting Capacities at Zero Degree Boom Angle
On Rubber - Defined Arc Over Front

Boom Angle	Main Boom Length in Feet				
	36	40	50	*60	70
0°	17,600 (29.7)	13,600 (34.2)	7,750 (44.2)	4,010 (54.6)	2,670 (64.2)

Note: ( ) Reference radii in feet. A6-829-101047

#LMI operating code. Refer to LMI manual for instructions.
*60 ft. boom length is with inner-mid extended and outer-mid & fly retracted.

Figure 18 Capacity differences.

of the crane, there may be load charts for outriggers at no extension, intermediate extension, and full extension (*Figure 20*, *Figure 21*, and *Figure 22*). For example, assume a boom length of 85 feet and an operating radius of 30 feet. The capacity for this configuration is shown on each of the three figures. At a 0-percent extension (*Figure 20*), the capacity is 13,650 pounds. At 50-percent extension (*Figure 21*), the capacity increases to 29,250 pounds—more than double the capacity at 0-percent extension. At 100-percent extension (*Figure 22*), the capacity increases to 36,700 pounds. Note that the minimum boom angle is unchanged in all three examples.

2.4.3 On Crawlers

Most crawler cranes include a feature for extending and retracting the crawler assemblies. Load chart capacities do not exceed 75 percent of tipping for each configuration on crawler cranes. Some cranes, however, may have outrigger extensions along with crawlers where 85 percent of tipping is applied.

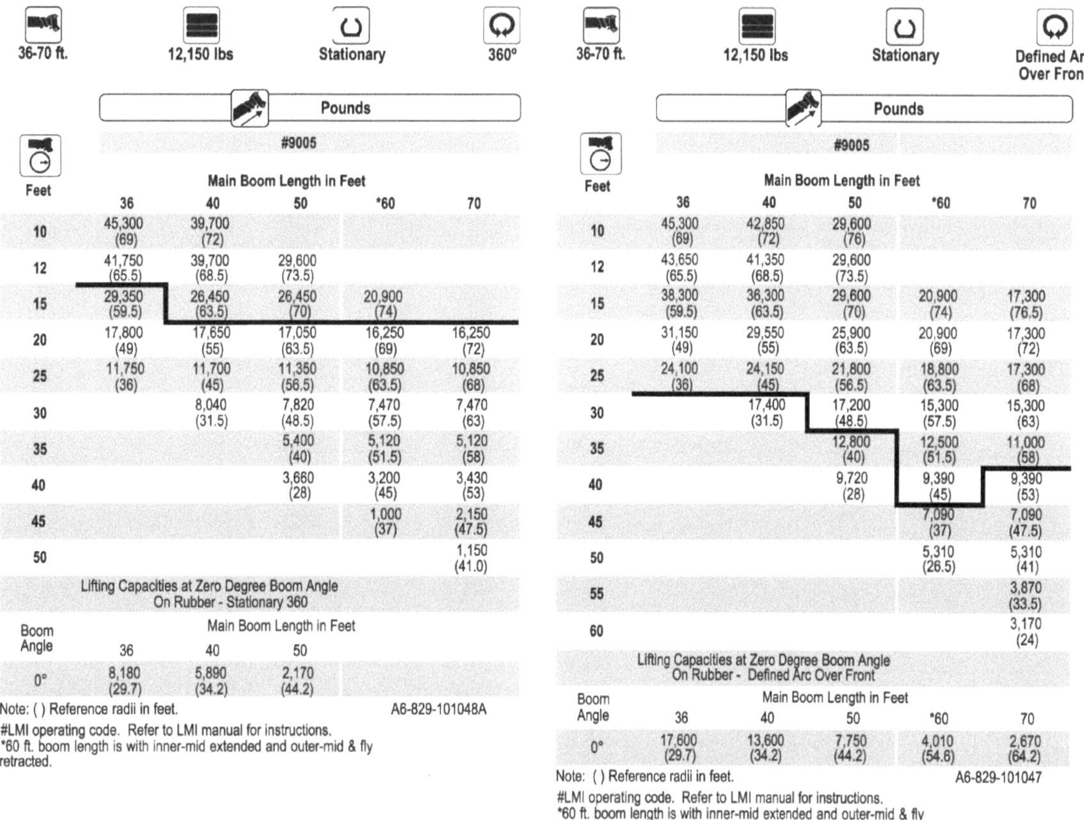

Pounds

#9005

On Rubber – Stationary 360

Feet	Main Boom Length in Feet				
	36	40	50	*60	70
10	45,300 (69)	39,700 (72)			
12	41,750 (65.5)	39,700 (68.5)	29,600 (73.5)		
15	29,350 (59.5)	26,450 (63.5)	26,450 (70)	20,900 (74)	
20	17,800 (49)	17,650 (55)	17,050 (63.5)	16,250 (69)	16,250 (72)
25	11,750 (36)	11,700 (45)	11,350 (56.5)	10,850 (63.5)	10,850 (68)
30		8,040 (31.5)	7,820 (48.5)	7,470 (57.5)	7,470 (63)
35			5,400 (40)	5,120 (51.5)	5,120 (58)
40			3,660 (28)	3,200 (45)	3,430 (53)
45				1,000 (37)	2,150 (47.5)
50					1,150 (41.0)

**Lifting Capacities at Zero Degree Boom Angle
On Rubber - Stationary 360**

Boom Angle	Main Boom Length in Feet		
	36	40	50
0°	8,180 (29.7)	5,890 (34.2)	2,170 (44.2)

Note: ( ) Reference radii in feet. A6-829-101048A

#LMI operating code. Refer to LMI manual for instructions.
*60 ft. boom length is with inner-mid extended and outer-mid & fly retracted.

Pounds

#9005

On Rubber – Defined Arc Over Front

Feet	Main Boom Length in Feet				
	36	40	50	*60	70
10	45,300 (69)	42,850 (72)	29,600 (76)		
12	43,650 (65.5)	41,350 (68.5)	29,600 (73.5)		
15	38,300 (59.5)	36,300 (63.5)	29,600 (70)	20,900 (74)	17,300 (76.5)
20	31,150 (49)	29,550 (55)	25,900 (63.5)	20,900 (69)	17,300 (72)
25	24,100 (36)	24,150 (45)	21,800 (56.5)	18,800 (63.5)	17,300 (68)
30		17,400 (31.5)	17,200 (48.5)	15,300 (57.5)	15,300 (63)
35			12,800 (40)	12,500 (51.5)	11,000 (58)
40			9,720 (28)	9,390 (45)	9,390 (53)
45				7,090 (37)	7,090 (47.5)
50				5,310 (26.5)	5,310 (41)
55					3,870 (33.5)
60					3,170 (24)

**Lifting Capacities at Zero Degree Boom Angle
On Rubber - Defined Arc Over Front**

Boom Angle	Main Boom Length in Feet				
	36	40	50	*60	70
0°	17,600 (29.7)	13,600 (34.2)	7,750 (44.2)	4,010 (54.6)	2,670 (64.2)

Note: ( ) Reference radii in feet. A6-829-101047

#LMI operating code. Refer to LMI manual for instructions.
*60 ft. boom length is with inner-mid extended and outer-mid & fly retracted.

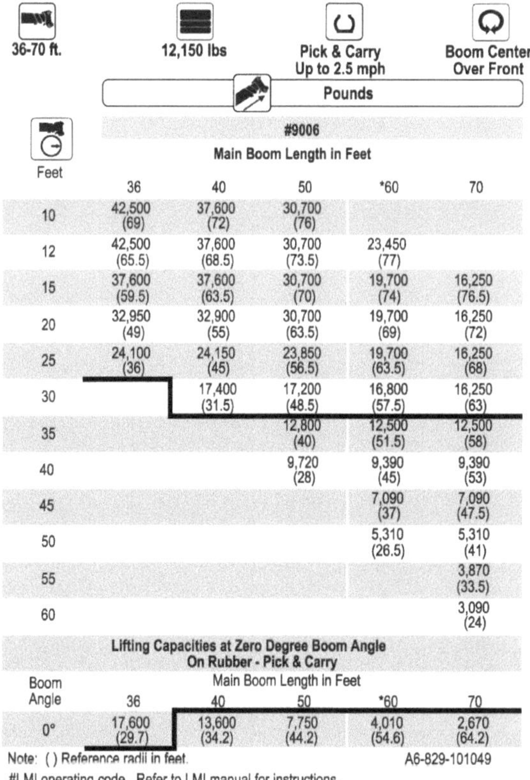

Pounds

#9006

Feet	Main Boom Length in Feet				
	36	40	50	*60	70
10	42,500 (69)	37,600 (72)	30,700 (76)		
12	42,500 (65.5)	37,600 (68.5)	30,700 (73.5)	23,450 (77)	
15	37,600 (59.5)	37,600 (63.5)	30,700 (70)	19,700 (74)	16,250 (76.5)
20	32,950 (49)	32,900 (55)	30,700 (63.5)	19,700 (69)	16,250 (72)
25	24,100 (36)	24,150 (45)	23,850 (56.5)	19,700 (63.5)	16,250 (68)
30		17,400 (31.5)	17,200 (48.5)	16,800 (57.5)	16,250 (63)
35			12,800 (40)	12,500 (51.5)	12,500 (58)
40			9,720 (28)	9,390 (45)	9,390 (53)
45				7,090 (37)	7,090 (47.5)
50				5,310 (26.5)	5,310 (41)
55					3,870 (33.5)
60					3,090 (24)

**Lifting Capacities at Zero Degree Boom Angle
On Rubber - Pick & Carry**

Boom Angle	Main Boom Length in Feet				
	36	40	50	*60	70
0°	17,600 (29.7)	13,600 (34.2)	7,750 (44.2)	4,010 (54.6)	2,670 (64.2)

Note: ( ) Reference radii in feet. A6-829-101049

#LMI operating code. Refer to LMI manual for instructions.
*60 ft. boom length is with inner-mid extended and outer-mid & fly retracted.

Figure 19 On-rubber load chart.

NOTES:

1. Capacities are in pounds and do not exceed 75% of tipping loads as determined by test in accordance with SAE J765.
2. Capacities are applicable to machines equipped with 29.6x25 (28 ply) tires at 65 psi cold inflation pressure.
3. Defined Arc – Over front includes 6' on either side of longitudinal centerline of machine (ref. drawing C6-829-003529).
4. Capacities appearing above the bold line are based on structural strength and tipping should not be relied upon as a capacity limitation.
5. Capacities are applicable only with machine on firm level surface.
6. On rubber lifting with boom extensions not permitted.
7. For pick and carry operation, boom must be centered over front of machine, mechanical swing lock engaged and load restrained from swinging. When handling loads in the structural range with capacities close to maximum ratings, travel should be reduced to creep speeds.
8. Axle lockouts must be functioning when lifting on rubber.
9. All lifting depends on proper tire inflation, capacity and condition. Capacities must be reduced for lower tire inflation pressures. See lifting capacity chart for tire used. Damaged tires are hazardous to safe operation of crane.
10. Creep – Not over 200 ft. of movement in any 30 minute period and not exceeding 1 mph.

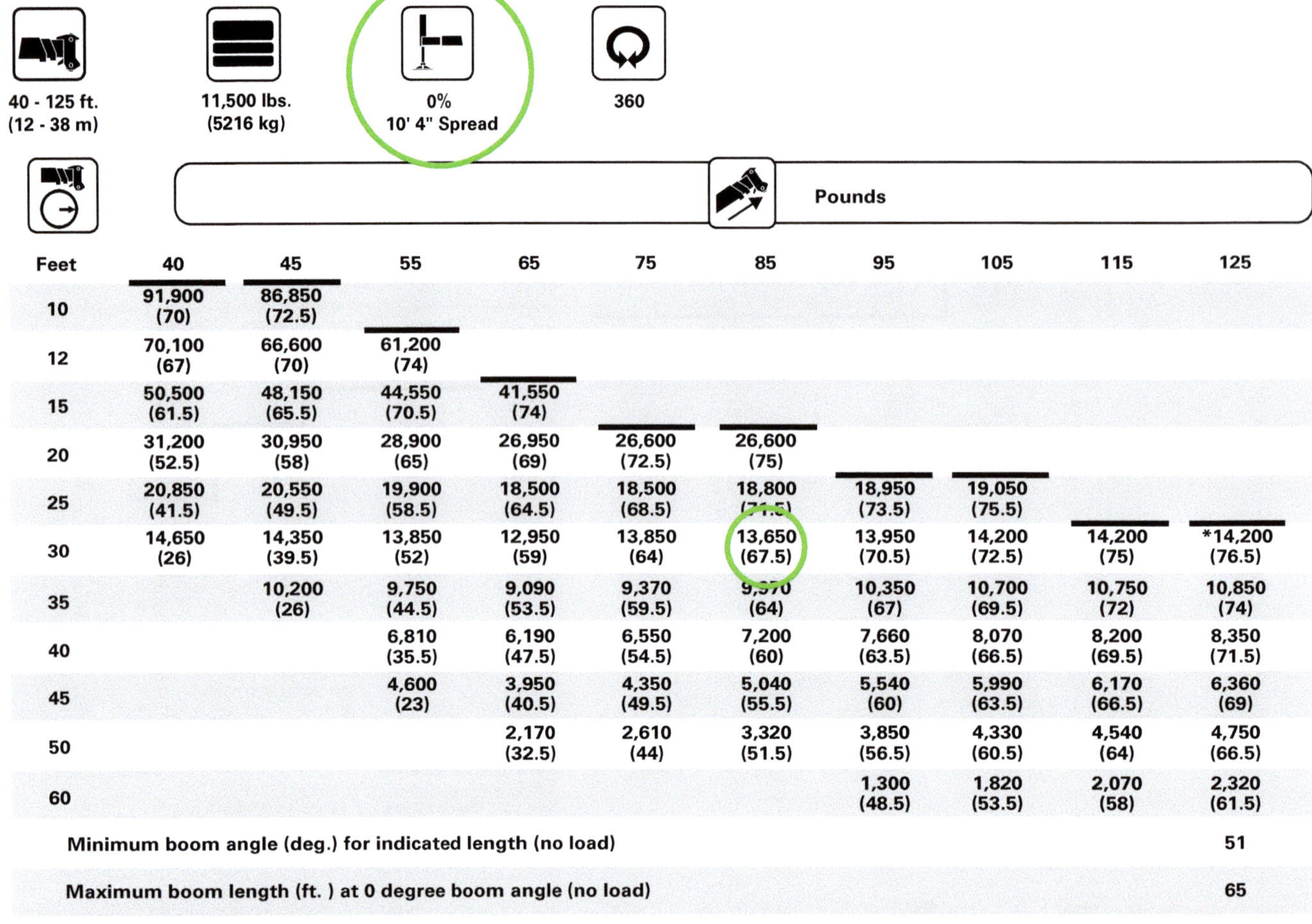

Feet	40	45	55	65	75	85	95	105	115	125
10	91,900 (70)	86,850 (72.5)								
12	70,100 (67)	66,600 (70)	61,200 (74)							
15	50,500 (61.5)	48,150 (65.5)	44,550 (70.5)	41,550 (74)						
20	31,200 (52.5)	30,950 (58)	28,900 (65)	26,950 (69)	26,600 (72.5)	26,600 (75)				
25	20,850 (41.5)	20,550 (49.5)	19,900 (58.5)	18,500 (64.5)	18,500 (68.5)	18,800 (71.5)	18,950 (73.5)	19,050 (75.5)		
30	14,650 (26)	14,350 (39.5)	13,850 (52)	12,950 (59)	13,850 (64)	13,650 (67.5)	13,950 (70.5)	14,200 (72.5)	14,200 (75)	*14,200 (76.5)
35		10,200 (26)	9,750 (44.5)	9,090 (53.5)	9,370 (59.5)	9,370 (64)	10,350 (67)	10,700 (69.5)	10,750 (72)	10,850 (74)
40			6,810 (35.5)	6,190 (47.5)	6,550 (54.5)	7,200 (60)	7,660 (63.5)	8,070 (66.5)	8,200 (69.5)	8,350 (71.5)
45			4,600 (23)	3,950 (40.5)	4,350 (49.5)	5,040 (55.5)	5,540 (60)	5,990 (63.5)	6,170 (66.5)	6,360 (69)
50				2,170 (32.5)	2,610 (44)	3,320 (51.5)	3,850 (56.5)	4,330 (60.5)	4,540 (64)	4,750 (66.5)
60							1,300 (48.5)	1,820 (53.5)	2,070 (58)	2,320 (61.5)

Minimum boom angle (deg.) for indicated length (no load)	51
Maximum boom length (ft.) at 0 degree boom angle (no load)	65

NOTE: ( ) Boom angles are in degrees.
*Based on maximum obtainable boom angle.

A6-829-011927A

Boom Angle	40	45	55
0	12,550 (32.3)	8,440 (37.8)	3,610 (47.8)

NOTE: ( ) Reference radii are in feet.

A6-829-012282

Figure 20 Capacities on outriggers at no extension (0 percent).

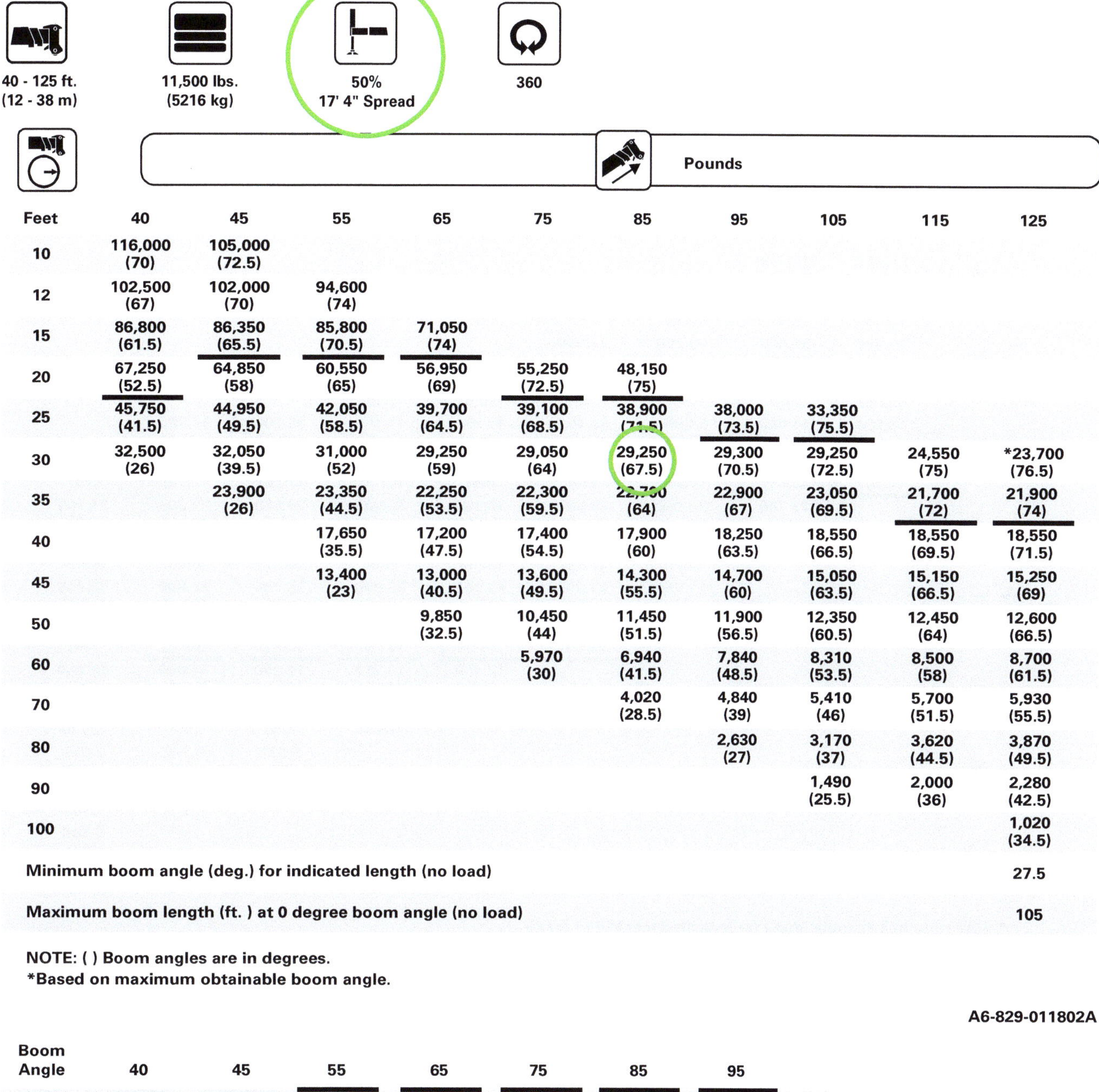

Feet	40	45	55	65	75	85	95	105	115	125
10	116,000 (70)	105,000 (72.5)								
12	102,500 (67)	102,000 (70)	94,600 (74)							
15	86,800 (61.5)	86,350 (65.5)	85,800 (70.5)	71,050 (74)						
20	67,250 (52.5)	64,850 (58)	60,550 (65)	56,950 (69)	55,250 (72.5)	48,150 (75)				
25	45,750 (41.5)	44,950 (49.5)	42,050 (58.5)	39,700 (64.5)	39,100 (68.5)	38,900 (71.5)	38,000 (73.5)	33,350 (75.5)		
30	32,500 (26)	32,050 (39.5)	31,000 (52)	29,250 (59)	29,050 (64)	29,250 (67.5)	29,300 (70.5)	29,250 (72.5)	24,550 (75)	*23,700 (76.5)
35		23,900 (26)	23,350 (44.5)	22,250 (53.5)	22,300 (59.5)	22,700 (64)	22,900 (67)	23,050 (69.5)	21,700 (72)	21,900 (74)
40			17,650 (35.5)	17,200 (47.5)	17,400 (54.5)	17,900 (60)	18,250 (63.5)	18,550 (66.5)	18,550 (69.5)	18,550 (71.5)
45			13,400 (23)	13,000 (40.5)	13,600 (49.5)	14,300 (55.5)	14,700 (60)	15,050 (63.5)	15,150 (66.5)	15,250 (69)
50				9,850 (32.5)	10,450 (44)	11,450 (51.5)	11,900 (56.5)	12,350 (60.5)	12,450 (64)	12,600 (66.5)
60					5,970 (30)	6,940 (41.5)	7,840 (48.5)	8,310 (53.5)	8,500 (58)	8,700 (61.5)
70						4,020 (28.5)	4,840 (39)	5,410 (46)	5,700 (51.5)	5,930 (55.5)
80							2,630 (27)	3,170 (37)	3,620 (44.5)	3,870 (49.5)
90								1,490 (25.5)	2,000 (36)	2,280 (42.5)
100										1,020 (34.5)

Minimum boom angle (deg.) for indicated length (no load) — 27.5

Maximum boom length (ft.) at 0 degree boom angle (no load) — 105

NOTE: ( ) Boom angles are in degrees.
*Based on maximum obtainable boom angle.

A6-829-011802A

Boom Angle	40	45	55	65	75	85	95
0	22,800 (32.3)	18,250 (37.8)	11,500 (47.8)	6,220 (57.8)	3,740 (67.8)	2,350 (77.8)	1,300 (87.8)

NOTE: ( ) Reference radii are in feet.

A6-829-012282

Figure 21 Capacities on outriggers at intermediate extension (50 percent).

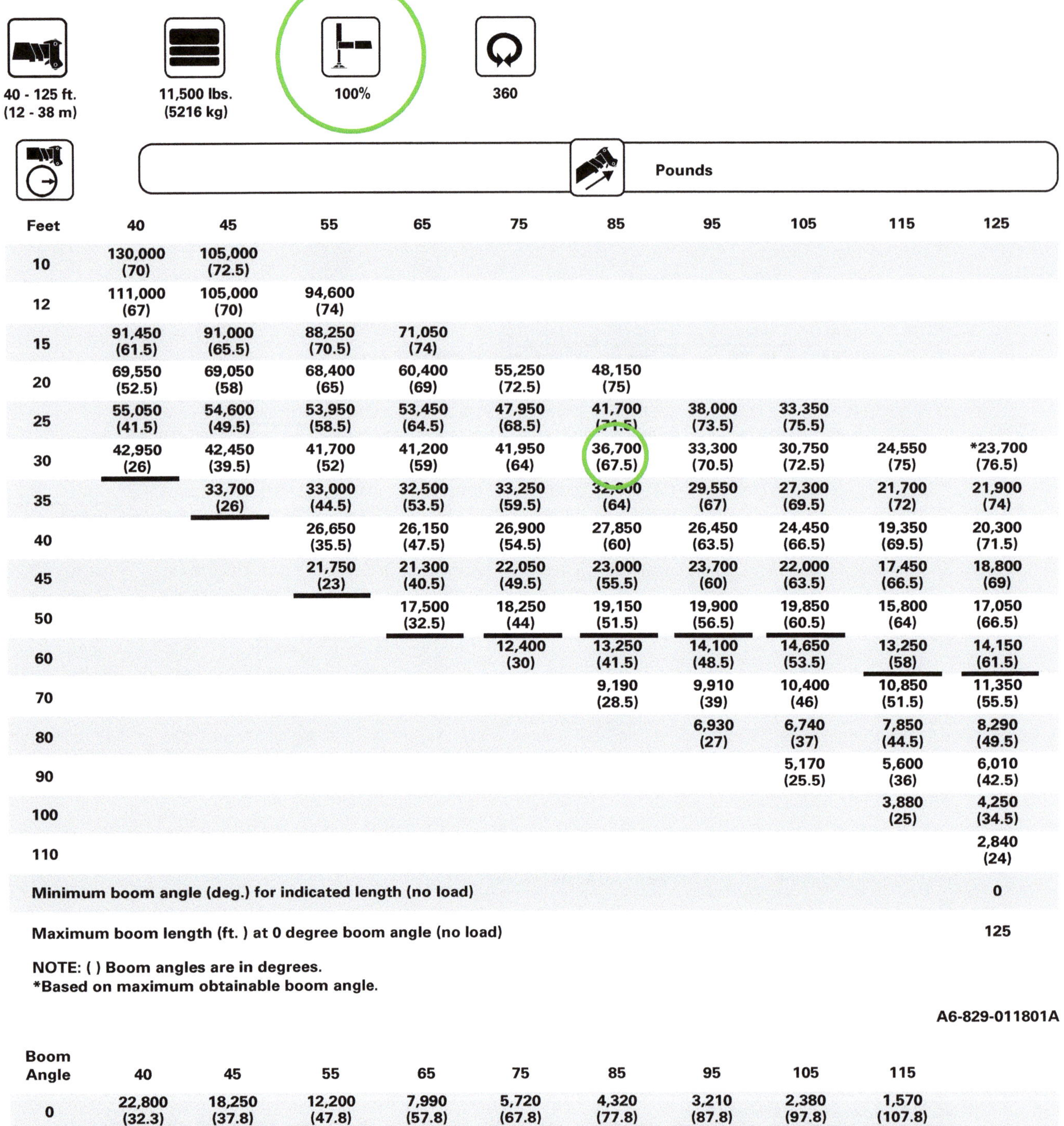

Pounds

Feet	40	45	55	65	75	85	95	105	115	125
10	130,000 (70)	105,000 (72.5)								
12	111,000 (67)	105,000 (70)	94,600 (74)							
15	91,450 (61.5)	91,000 (65.5)	88,250 (70.5)	71,050 (74)						
20	69,550 (52.5)	69,050 (58)	68,400 (65)	60,400 (69)	55,250 (72.5)	48,150 (75)				
25	55,050 (41.5)	54,600 (49.5)	53,950 (58.5)	53,450 (64.5)	47,950 (68.5)	41,700 (71.5)	38,000 (73.5)	33,350 (75.5)		
30	42,950 (26)	42,450 (39.5)	41,700 (52)	41,200 (59)	41,950 (64)	36,700 (67.5)	33,300 (70.5)	30,750 (72.5)	24,550 (75)	*23,700 (76.5)
35		33,700 (26)	33,000 (44.5)	32,500 (53.5)	33,250 (59.5)	32,000 (64)	29,550 (67)	27,300 (69.5)	21,700 (72)	21,900 (74)
40			26,650 (35.5)	26,150 (47.5)	26,900 (54.5)	27,850 (60)	26,450 (63.5)	24,450 (66.5)	19,350 (69.5)	20,300 (71.5)
45			21,750 (23)	21,300 (40.5)	22,050 (49.5)	23,000 (55.5)	23,700 (60)	22,000 (63.5)	17,450 (66.5)	18,800 (69)
50				17,500 (32.5)	18,250 (44)	19,150 (51.5)	19,900 (56.5)	19,850 (60.5)	15,800 (64)	17,050 (66.5)
60					12,400 (30)	13,250 (41.5)	14,100 (48.5)	14,650 (53.5)	13,250 (58)	14,150 (61.5)
70						9,190 (28.5)	9,910 (39)	10,400 (46)	10,850 (51.5)	11,350 (55.5)
80							6,930 (27)	6,740 (37)	7,850 (44.5)	8,290 (49.5)
90								5,170 (25.5)	5,600 (36)	6,010 (42.5)
100									3,880 (25)	4,250 (34.5)
110										2,840 (24)

Minimum boom angle (deg.) for indicated length (no load)	0
Maximum boom length (ft.) at 0 degree boom angle (no load)	125

NOTE: ( ) Boom angles are in degrees.
*Based on maximum obtainable boom angle.

A6-829-011801A

Boom Angle	40	45	55	65	75	85	95	105	115
0	22,800 (32.3)	18,250 (37.8)	12,200 (47.8)	7,990 (57.8)	5,720 (67.8)	4,320 (77.8)	3,210 (87.8)	2,380 (97.8)	1,570 (107.8)

NOTE: ( ) Reference radii are in feet.

A6-829-012282

Figure 22 Capacities on outriggers at full extension (100 percent).

2.4.4 Tower and Ring Attachments

A tower luffing-jib attachment (*Figure 23*) is an alternative to a freestanding tower crane where a lower erection height allows its use. It can also increase the load radius of a standard crane where reach is the primary concern. As with all attachments, the operator should follow the crane manufacturer's and/or attachment manufacturer's recommendations for lifting.

When a work site requires a stationary crane with major capacity, some manufacturers offer ringer crane (*Figure 24*) packages. A ring attachment installed on the ground greatly increases the crane's capacity and stability, permitting weight transfer to the ring foundation in all positions. The capacity remains the same regardless of the operating quadrant.

2.4.5 Counterweight Configurations

Counterweights give extra stability to the crane during assembly and lifting operations. A crane can have front, rear, side, or upper counterweights. Each crane has counterweight specifications, and manufacturers explain in the operational notes that the capacities listed are with the specified counterweights in place. The assembled counterweight sections have to match the configurations specified in the load charts. If a counterweight is extendable, its extension must be according to the manufacturer's recommendations. *Figure 25* and *Figure 26* show examples of counterweight configurations.

It's important to note that no form of counterweight can be used other than the counterweight specified by the manufacturer. For example, placing or connecting another heavy piece of equipment onto the crane is prohibited.

Figure 24 3,000 metric-ton ringer crane.

Figure 23 Crane with tower attachment.

Figure 25 Crawler crane counterweights.

Figure 26 Counterweights on an auxiliary carrier.

Additional Resources

ASME Standard B30.5, Mobile and Locomotive Cranes. Current edition. New York, NY: American Society of Mechanical Engineers.

Cranes: Design, Practice, and Maintenance, Ing J. Verschoof. 2002. Hoboken, NJ: John Wiley and Sons, Inc.

29 *CFR* 1926, Subpart CC, *Cranes and Derricks in Construction*. **www.ecfr.gov**

2.0.0   Section Review

1. Relatively small four-wheeled cranes typically operated by the driver are _____.

   a. industrial/all-purpose cranes
   b. crawler cranes
   c. lattice-boom cranes
   d. articulated boom cranes

2. The arrangement of information on a crane load chart _____.

   a. is specified by safety standards
   b. varies according to the manufacturer and crane type
   c. always shows boom height along the top row
   d. always shows gross capacities in the right column

3. Most mobile cranes are more stable front-to-rear because _____.

   a. the crane upperworks are centered on the carrier chassis
   b. they are typically wider than they are long
   c. the carrier's drive engine lowers the center of gravity
   d. they are typically longer than they are wide

4. All the items listed here relate specifically to the crane base configuration for load chart purposes *except* _____.

   a. fuel-tank fill level
   b. outriggers
   c. tires
   d. crawlers

3.0.0 Working with Load Charts

Objective

Describe how to apply load charts and deductions from capacity.

   a. Describe the use of jib charts.
   b. Describe how the parts of line can affect capacity.
   c. Describe how to read a range diagram.
   d. Describe how to calculate a crane's capacity from its load charts.

Performance Tasks

4. Properly identify load charts that are used in different configurations.
5. Identify parts of line and counterweight considerations in load chart information.
6. Calculate minimum parts of line required.

Trade Terms

Effective weight: With reference to jib/boom extensions, the weight calculated by the manufacturer which, when applied at the boom tip, has the same effect on the crane capacity as the jib and/or boom extension itself.

Offset: The angle formed by an erected boom, measured from an imaginary line extending through the center line of the main boom. The crane's load chart provides capacities for allowable jib or boom-extension offset angles.

Line pull: The maximum pulling force a crane hoist drum can apply to a hoist cable, accounted for by the torque that the motor can apply to the drum and the working radius of the rope as it leaves the drum.

Load charts establish the gross weight that a crane can lift in a specific configuration. Extra weight, such as the load block, headache ball, wire rope, and parts of hoist line, reduce the primary load weight that the crane can safely hoist. These weights must be deducted from the gross capacity of the crane unless they are specifically included in the load chart configuration. The weight of the main boom is configured and included.

Load charts usually have a separate list of items with their weight deductions (*Figure 27*). Also refer to the load charts in *Appendix B* for examples of these lists. Operators should always review the notes section of load charts as well. The notes provide valuable information that must often be applied.

Some load charts require the deduction of the suspended hoist line, while others include the rope weight in the load chart calculations. If you must calculate a rope deduction, find the product of (multiply) the full boom-tip height from which the load is suspended, the parts of line, and the line's weight per unit. This is illustrated by the following equation:

Suspended line weight (lb) =
   boom-tip height (ft) × parts of line × unit weight (lb/ft)

Assume that the boom-tip height of a crane is 45 feet, there are 3 parts of line, and the ¾" wire rope weighs 1.15 lb/ft:

Suspended line weight (lb) =
   45 ft × 3 parts of line × 1.15 lb/ft = **156 lb**

The exact result of this equation is actually 155.25 lb. However, when determining the weight of variable loads like rope, these calculations are usually rounded up to the nearest whole number (in this case, 156 lb). This calculation can also be used with metric units, substituting kilograms and meters, for example. It provides a conservative value for the weight of suspended cable, which can amount to thousands of pounds for larger cranes.

3.1.0 Jib Charts

Jib load charts, such as the one shown in *Figure 28*, show the capacities in pounds, kips, or kilograms for the luffing or fixed boom extensions at different degrees of jib offset.

> **NOTE**
>
> Jib chart styles vary widely from manufacturer to manufacturer.

The following factors affect the capacity of a crane equipped with a jib boom extension:

- Load radius
- Length of the main boom
- Main boom angle
- Jib length
- Jib angle with respect to the boom (offset)
- Weight of the rigging and hooks on both the jib and main boom

Load Handling Equipment:	(lbs)
Auxiliary Head Attached	100
25-ton quick reeve 3 sheave hook block (see hook block for actual weight)	670
40-ton quick reeve 4 sheave hook block (see hook block for actual weight)	780
8.5-ton hook ball (see hook ball for actual weight)	360
Lifting From Main Boom With:	(lbs)
28.5 ft or 51 ft fly stowed on base (see operation note 4)	0
28.5 ft. offset fly erected but not used	2600
51 ft. offset fly erected but not used	4800
Lifting From 28.5 ft Offset Fly With:	
22.5 ft fly tip erected but not used	**PROHIBITED**
22.5 ft. fly tip stowed on 28.5 ft. offset fly	**PROHIBITED**
Note: Capacity deductions are for Link-Belt supplied equipment <u>only</u>.	

Figure 27 Deductions chart.

When not in use, some jibs and/or boom extensions are stowed on the main boom. In some cases, the operator must deduct the extension's weight from the crane's gross capacity in this configuration. The operator must consult the manufacturer's documentation to determine the effect of a stowed jib/boom extension on the crane's net capacity.

When mounted on the boom tip, the operator must deduct the **effective weight** of the jib/extension from the crane's gross capacity. This deduction is made only when hoisting with the main boom. When lifting from an erected jib, do not count the weight of the jib as part of the load; the appropriate load chart already compensates for the specified jib configuration.

3.2.0 Line Pull and Parts of Line

Determination of a crane's gross and net capacity using a load chart assures the operator that the crane configuration will allow hoisting a given load. However, to physically lift a load requires pulling on the hoist line. How do you know if the line is strong enough to lift the weight of the load, and whether the crane can apply enough tension to the hoist line?

3.2.1 Hoist Rope Properties

Crane hoist rope must be strong enough for its intended purpose, flexible enough to run smoothly through sheaves, and have minimal tendency to rotate or twist under load. Unfortunately, these properties can work against each other. Rope manufacturers compromise to obtain the optimum rope qualities for an intended purpose. Strength and flexibility are obtained by using the finest wire strands possible and plying them in various ways to maximize flexibility at a given strength.

Rotation-resistant wire rope is manufactured by applying multiple strand layers in opposite directions. Thus, when the rope is under load, it resists untwisting and spinning the load. While rotation-resistant rope provides this desirable characteristic, it is not as durable as its standard counterpart.

For safety purposes, a wire rope's breaking strength must be much larger than any tension it will experience during operation. *Table 1* lists typical wire rope sizes, associated breaking strengths, and their weights for a given type of rotation-resistant rope.

Table 1 Selected Wire Rope Properties

Wire Rope Diameter (in)	Minimum Breaking Strength (lb)	Unit Weight (lb/ft)
$^5/_8$	47,400	0.78
$^3/_4$	69,400	1.15
1	122,800	2.09
$1^1/_2$	269,000	4.56

33-56 ft. luffing folding boom extension (mode B) (fixed offset angles)

37.3-141.7 ft. | 33 - 56 ft. | 22,000 lbs | 100% 34'-6" Spread | 360°

Feet	33 ft. LENGTH			56 ft. LENGTH		
	5° OFFSET #0091	20° OFFSET #0091	40° OFFSET #0091	5° OFFSET #0092	20° OFFSET #0092	40° OFFSET #0092
40	*13,700 (78)					
45	13,700 (77)					
50	13,700 (75)	13,700 (77.5)		*8,200 (78)		
55	13,700 (73.5)	13,700 (75.5)	*11,000 (78)	8,200 (77.5)		
60	13,700 (71.5)	13,700 (74)	11,000 (76)	8,200 (76)		
65	13,700 (70)	12,850 (72)	10,950 (74.5)	8,200 (74.5)	8,200 (77.5)	
70	12,500 (68)	12,000 (70)	10,350 (72.5)	8,200 (73)	8,200 (76)	
75	11,350 (66)	11,200 (68)	9,830 (70.5)	8,200 (71.5)	8,100 (74)	6,400 (77.5)
80	9,730 (64.5)	10,450 (66.5)	9,330 (68.5)	8,200 (69.5)	7,600 (72.5)	6,400 (76)
85	8,300 (62.5)	8,980 (64.5)	8,860 (66.5)	8,200 (68)	7,150 (71)	6,230 (74)
90	7,060 (60.5)	7,660 (62.5)	8,210 (64.5)	7,740 (66.5)	6,730 (69)	5,920 (72.5)
95	5,960 (58.5)	6,500 (60.5)	6,980 (62)	7,130 (64.5)	6,350 (67.5)	5,640 (70.5)
100	4,990 (56.5)	5,470 (58)	5,880 (60)	6,130 (63)	6,000 (65.5)	5,380 (68.5)
105	4,120 (54)	4,560 (56)	4,900 (58)	5,230 (61)	5,690 (64)	5,140 (67)
110	3,340 (52)	3,730 (54)	4,020 (55.5)	4,430 (59.5)	5,290 (62)	4,900 (65)
115	2,640 (49.5)	2,990 (51.5)	3,230 (53)	3,700 (57.5)	4,490 (60)	4,690 (63)
120	2,000 (47.5)	2,320 (49)	2,510 (50.5)	3,040 (55.5)	3,760 (58.5)	4,470 (61)
125	1,420 (45)	1,700 (46.5)	1,850 (47.5)	2,440 (53.5)	3,100 (56.5)	3,710 (58.5)
130		1,140 (44)	1,250 (45)	1,900 (51.5)	2,500 (54.5)	3,030 (56.5)
135				1,390 (49.5)	1,940 (52)	2,390 (54)
140					1,420 (50)	1,810 (52)
145						1,270 (49)
Minimum boom angle (°) for indicated length (no load)	42	43	43	48	48	47
Maximum boom length (ft.) at 0° boom angle (no load)		89.8			76.7	

Pounds

NOTE: ( ) Boom angles are in degrees.

A6-829-103522

#LMI operating code. Refer to LMI manual for operating instructions.
*This capacity is based upon maximum boom angle.

NOTES:

1. All capacities above the bold line are based on structural strength of boom extension and do not exceed 85% of tipping loads, in accordance with SAE J-765.

2. The 33 ft. luffing folding boom extension may be used for single or double line lifting service. The 56 ft. luffing folding boom extension may be used for single line lifting service only.
   WARNING: Lifting with the 33 ft. extension base, with the 23 ft. extension fly either erected or folded along side of extension base, is strictly prohibited.

3. WARNING: Operation of this machine with heavier loads than the capacities listed is strictly prohibited. Machine tipping with boom extension occurs rapidly and without advance warning.

4. Boom angle is the angle above or below horizontal of the longitudinal axis of the boom base section after lifting rated load.

5. Capacities listed are with outriggers properly extended and vertical jacks set only.

6. For main boom lengths less than 141.7 ft. with the boom extension erected, the rated loads are determined by boom angle. Use only the column which corresponds to the boom extension length and offset for which the machine is set up. For boom angles not shown, use rating of the next lower boom angle.

7. When lifting over the main boom nose with 33 ft. or 56 ft. extension erected, the outriggers must be fully extended or 50% extended (17.3 ft. spread).

Figure 28 Jib/boom extension chart.

3.2.2 Line Pull

Manufacturers describe a crane's ability to apply tension to the hoist cable as its line pull. This is the maximum amount of force the hoist drum can apply to the hoist cable for a given layer on the drum. A hoist drum motor can apply only a certain amount of torque to the drum. As the drum wraps more layers of rope, its diameter increases, and the maximum line pull it can exert decreases proportionately. It is an industry standard that the ratio of the minimum breaking strength of the specified hoist cable to the crane's maximum line pull should be at least 5:1. Therefore, the limiting factor is usually the crane's line pull, not the minimum breaking strength of the wire rope.

Another factor that affects line pull is the type of hoist sheave bearings the crane has. Cranes have sheaves with either bronze-bushed bearings or low-friction roller bearings. The type of bearings affects the amount of friction force that the line pull must overcome during the lift. Also, the number of sheaves and their location in the system affect how fast they are moving and the amount of friction they generate. To lift a certain load, the line pull must be greater than the suspended load weight, plus the additional resistance created by the friction force of the rotating sheaves.

3.2.3 Parts of Line

Maximum crane capacities may far exceed the breaking strength of a single hoist line. So how do cranes manage to lift such loads? By using multiple lines to support a load block, the weight of the load is divided among the number of line parts reeved through the block. Two parts of line reduce the weight each line supports by half. Each of five lines supports only one-fifth of the suspended load. Before making a lift, the operator must ensure the crane has the required minimum parts of line (*Figure 29*) to lift the load.

NOTE

The rope dead end of an odd number of parts attaches to the load block. The dead end of an even number of parts attaches to the boom head/sheave housing.

There are several methods the operator can use to determine the parts of line needed for a lift. Some crane manufacturers provide this information along with the load charts (*Figure 30*). This is the easiest method, and it is safe and accurate if the crane has the manufacturer's specified wire rope.

Without this resource, use the following procedure to determine the minimum number of parts of line required for a given lift:

Step 1 Determine the suspended weight of the load, rigging, and other hoisting accessories.

Step 2 Determine the crane's minimum single line pull available during the lift operation.

Step 3 Determine the type of hoist sheave bearings in the hoist system.

Step 4 Divide the suspended weight found in Step 1 by the crane's single line pull.

Step 5 Compare the value found in Step 4 to the listed ratios found in *Table 2* for either the bronze-bushed sheaves or the anti-friction or roller bearing sheaves. Always round up for intermediate ratios.

Step 6 In the Number of Line Parts column, read the number corresponding to the resulting ratio found in Step 5. This is the minimum required parts of line for the lift.

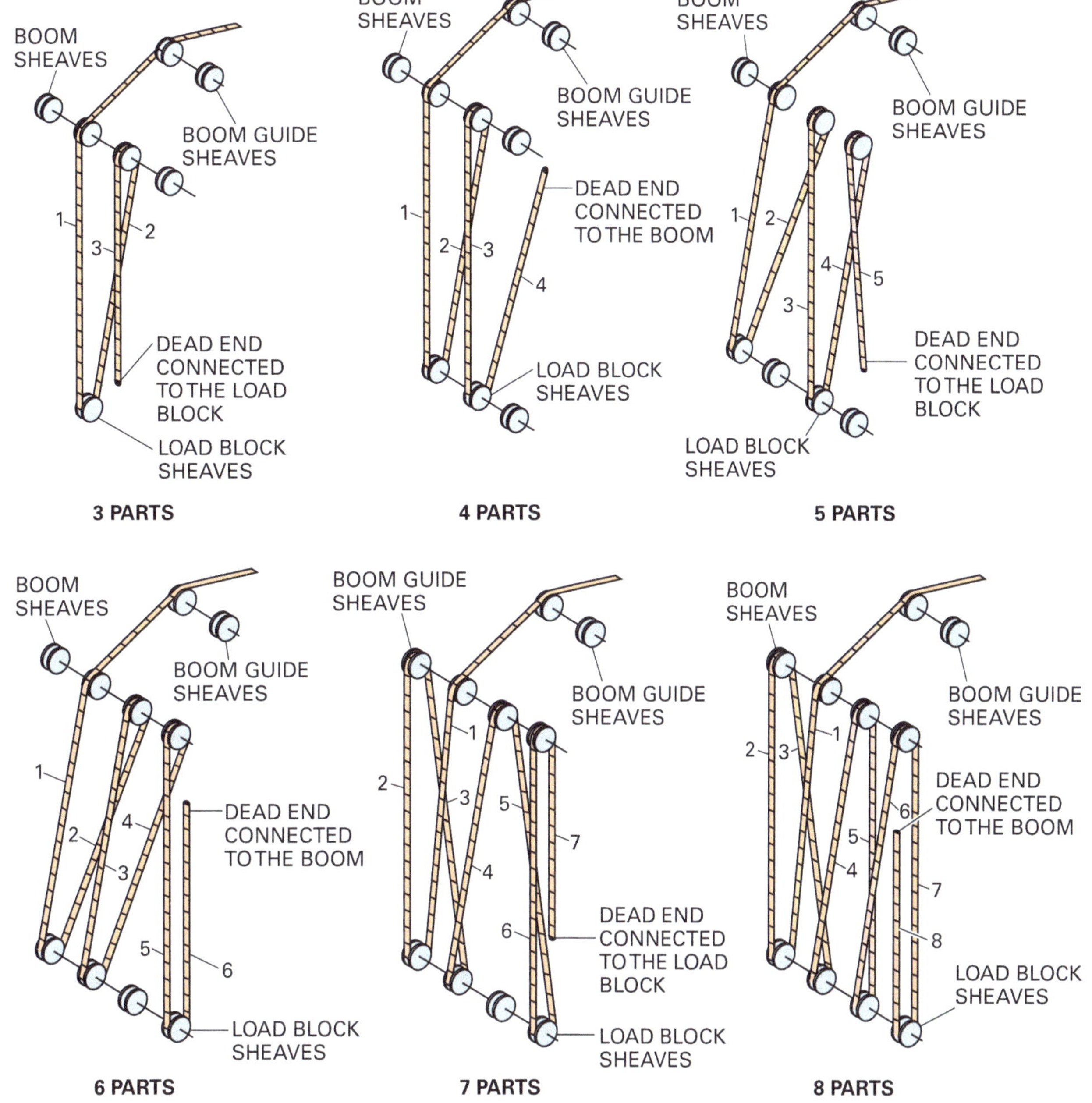

Figure 29 Reeving for various parts of line.

3.3.0 Range Diagram

A range diagram (*Figure 31*) is a side view of a crane with its full range of boom lengths, jibs, and boom extensions. Manufacturers include a range diagram in their load charts. It serves three important functions:

- It determines the configuration of the crane for a lift. The operator can determine the boom length, boom angle, boom point height or hook elevation, jib length, and jib offset required to suit the lift conditions at the site.

- It supports determining the main boom angle of telescopic booms for a given partial extension when only the load radius is known.
- It may identify the minimum allowable clearances between the load block and boom tip.

The operator should consult the range diagram at the beginning of the lift planning to establish the best configuration to conduct the lift. Planners must give consideration to the location of the load, where it needs to go, any site restrictions, and the dimensions and capabilities of the crane. The range diagram does not show crane lift capacities. After a configuration and basic setup has been chosen, the operator will consult the load chart to determine if the crane has the needed capacity in the proposed configuration.

3.4.0 Calculating Crane Capacity

There can be many factors involved in accurately determining a crane's capacity in a given configuration. This section explains, step by step, how to find the net capacity for each of the four crane categories.

Crane manufacturers provide operational notes with the load charts for each crane model. (Refer again to *Figure 28* for an example of these notes.) They provide very important information about the use of the chart as well as operating and safety guidance. The operator must read and understand these notes before making capacity selections.

Work through the following four examples and their accompanying load charts to become familiar with the factors involved in determining gross and net crane capacities. A standard method for determining the net capacity of a crane follows this simple pattern:

- Establish the gross capacity based on the planned configuration.
- Identify and sum all the deductions the configuration requires.
- Subtract all deductions from the gross capacity to determine net capacity.

WIRE ROPE CAPACITY

Maximum Lifting Capacities Based On Wire Rope Strength			
Parts of Line	5/8" Type RB	5/8" Type RB	Notes
1	9,080	11,080	Capacities shown are in pounds and working loads must not exceed the ratings on the capacity charts in the Crane Rating Manual.
2	18,160	22,160	
3	27,240	33,240	
4	36,320	44,320	
5	45,400	55,400	
6	54,480	66,480	Study Operator's Manual for wire rope inspection procedures and single part of line applications.
7	63,560	77,560	
8	72,640	88,640	
9	81,720		
LBCE	**DESCRIPTION**		
TYPE RB	18 × 19 Rotation Resistant – Compact Strand, High Strength Preformed, Right Regular Lay		
TYPE ZB	36 × 7 Rotation Resistant – Extra Improved Plow Steel – Right Regular Lay		

Figure 30 Wire rope capacity chart.

Table 2 Sheave Ratio Factors for Parts of Line

Ratio A: Bronze Bushed Sheaves	Ratio B: Anti-friction Bearing Sheaves	Number Of Line Parts
.96	.98	1
1.87	1.94	2
2.75	2.88	3
3.59	3.81	4
4.39	4.71	5
5.16	5.60	6
5.90	6.47	7
6.60	7.32	8
7.27	8.16	9
7.91	8.98	10
8.52	9.79	11
9.11	10.60	12
9.68	11.40	13
10.20	12.10	14
10.70	12.90	15
11.20	13.60	16
11.70	14.30	17
12.20	15.00	18
12.60	15.70	19
13.00	16.40	20

Link-Belt
CONSTRUCTION EQUIPMENT

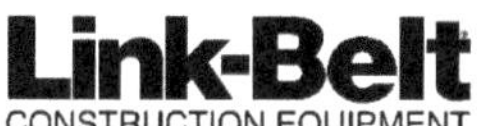

WORKING RANGE DIAGRAM

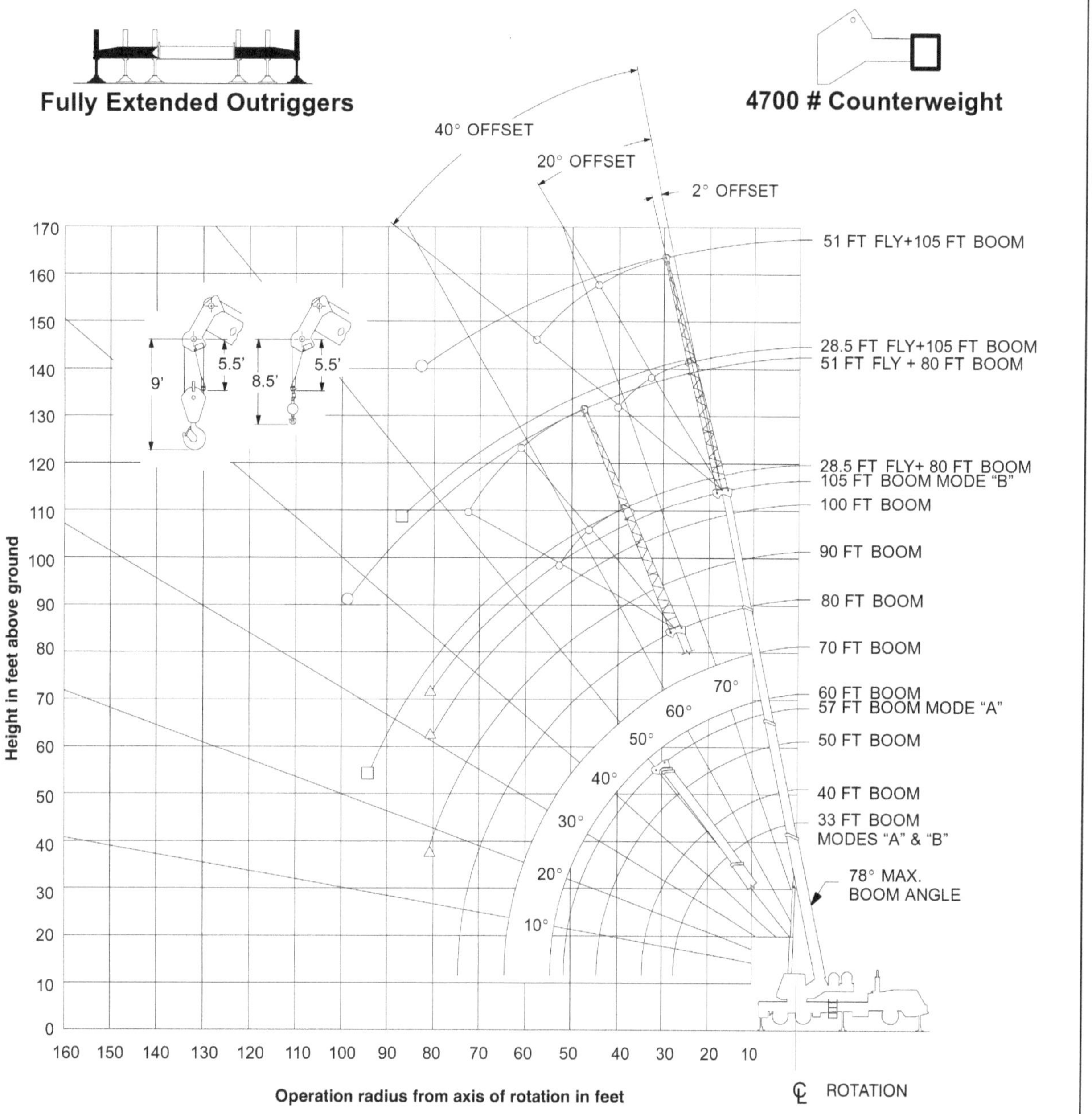

○ **Denotes Main Boom + 51 FT Offset Fly–Boom Mode "B"**

□ **Denotes Main Boom + 28.5 FT Offset Fly–Boom Mode "B"**

△ **Denotes Main Boom – Boom Mode "B"**

Note: Boom and fly geometry shown are for unloaded condition and crane standing level on firm supporting surface. Boom deflection, subsequent radius and boom angle change must be accounted for when applying load to hook.

WARNING

Do Not Lower The Boom Below The Minimum Boom Angle For No Load Stability As Shown In The Lift Charts For The Boom Lengths Given. Loss Of Stability Will Occur Causing A Tipping Condition.

Figure 31 Range diagram.

3.4.1 *Example One: Industrial/All-Purpose Crane*

Find the net capacity of a Grove YB4409 industrial crane (*Figure 32*) for a pick-and-carry operation, and its maximum allowable travel speed, configured as follows:

- On-rubber operation
- Over-front carry; boom centered
- 4-section main boom with fourth section retracted
- Load radius of 12 feet
- 6-foot jib stowed
- Equipped with manufacturer-specified tires at the specified pressure
- Hook and headache ball
- 40 lb of rigging

A simple graphic illustration of the configuration is shown in *Figure 33*.

The first step is to determine the crane's gross capacity in the planned configuration. Refer to the Grove YB4409 load chart in *Figure 34*. Verify that the crane's configuration agrees with the operating notes. Then, locate the column 4-Section Boom with 4th Retracted. Select the sub-column for On-Rubber, and its sub-column, F/R (Front/Rear). Read down the column to the row for the 12-foot load radius. The intersection of these two values indicates that the gross capacity for this crane is 3,400 lbs for over-front or over-rear operation.

Then, identify the required deductions from the gross capacity. Locate the deduction chart in the lower left corner of the load chart. Review the items on the chart and determine if any deductions are required. There is no load block in this configuration. The crane is equipped with a hook and ball, resulting in a deduction of 100 lb. The jib is stowed and the main boom is in use, so no deduction is required there.

Finally, determine the net capacity by subtracting any deductions from the gross capacity. You must deduct the headache ball (100 lb) and the rigging components (40 lb) listed in the configuration (*Figure 33*), for a total deduction of 140 lb (since 100 lb + 40 lb = 140 lb). The calculation is completed as follows:

$$\begin{array}{r} 3{,}400 \text{ lb gross capacity} \\ -\ 140 \text{ lb deductions} \\ \hline \mathbf{3{,}260 \text{ lb net capacity}} \end{array}$$

According to the deduction chart, no deduction is required for the weight of the hoist cable; its weight has been considered in the chart values. Note that a reeved load block is required to achieve the highest capacities for the crane. The operator should check the maximum line pull allowed for a single line. The rigging chart shown in *Figure 34* indicates that the maximum single-line capacity for the specified ½" rope is 8,500 lb, so the net capacity for this operation is well within this limit; a load block reeved with additional parts of line is not required for the 3,400 lb gross capacity the configuration allows.

Figure 32 Grove YB4409 industrial crane.

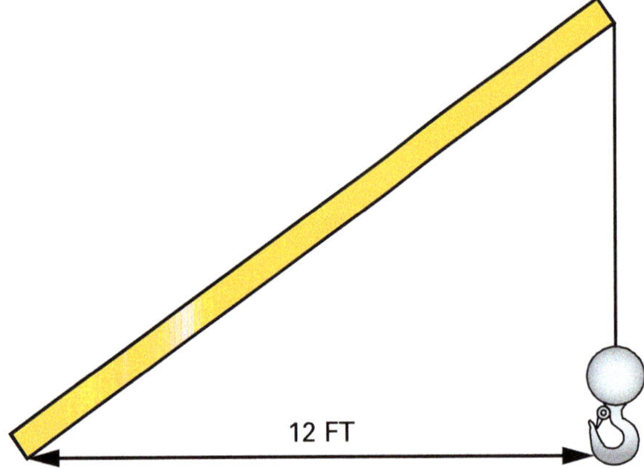

Figure 33 Graphic illustration of YB4409 configuration.

- ON-RUBBER OPERATION
- OVER-FRONT CARRY; BOOM CENTERED
- 4-SECTION MAIN BOOM WITH 4TH SECTION RETRACTED
- LOAD RADIUS OF 12 FEET
- 6-FOOT JIB STOWED
- TIRES AT THE SPECIFIED PRESSURE
- HOOK AND HEADACHE BALL
- 40 LB OF RIGGING

The maximum travel speed allowed during the operation is also found on the load chart. Note 5 in *Figure 34* specifies that loads on rubber may be transported at a maximum speed of 2.5 mph on a smooth surface, with the boom retracted to the shortest possible length and centered over the front.

3.4.2 Example Two: Telescopic-Boom Crane

When lifting from the main boom, find the net capacity of a Grove TM-1500 truck-mounted telescopic-boom crane (*Figure 35*) configured as follows:

- Outriggers properly set, fully extended, and level
- Over-the-rear lift
- 33'–58' extension stowed
- Boom length of 52 feet
- Load radius of 23 feet
- 150-ton, 8-sheave block
- Auxiliary boom head attached
- 10-ton headache ball reeved over auxiliary boom head
- Minimum reeving for the load weight will be used
- 195 lb of rigging and other hoisting accessories

A graphic illustration of the configuration is provided in *Figure 36*.

Again, the first step is to determine the gross capacity. Review the load chart notes shown in *Figure 37*. Per Note 2, no additional weight needs to be deducted for wire rope because the minimum reeving required for the load weight will be used. Also refer to Note 5, which indicates that if a dimension falls between the listed values, the next larger radius and/or boom length should be selected.

Study the load chart in *Figure 38*. Read down the Radius in Feet column to locate 23 ft. Since this specific radius isn't listed, go to the next longer load radius, which is 25 ft. Read across the row to the boom length column for 52 ft. This boom length isn't listed on the chart either, so go to the next longer boom length of 58 ft. The gross capacity where these two values intersect is 131,500 lb.

Now determine which deductions are appropriate for the configuration (refer to *Figure 39*). Include the 195 lb of rigging, which won't be shown on this chart. Adding the applicable deductions results in the following:

$$\begin{array}{r} 1{,}246 \text{ (33'–58' extension stowed)} \\ 5{,}254 \text{ (150-ton, 8-sheave block)} \\ 261 \text{ (auxiliary boom head)} \\ 560 \text{ (10-ton headache ball)} \\ +\quad 195 \text{ (rigging)} \\ \hline \textbf{7,516 lb total deduction} \end{array}$$

Finally, determine the net capacity by subtracting the deductions from the gross capacity, as shown here:

$$\begin{array}{r} 131{,}500 \text{ lb gross capacity} \\ -\quad 7{,}516 \text{ lb total deduction} \\ \hline \textbf{123,984 lb net capacity} \end{array}$$

3.4.3 Example Three: Lattice-Boom Crane

Find the net capacity of the Manitowoc 4100W-2 lattice-boom crawler crane (*Figure 40*) in the following configuration:

- Counterweights totaling 206,400 lb installed, including (2) 30,000-lb carbody counterweights and (2) 12,000-lb side counterweights
- Crawlers extended
- #22C boom, 230' long
- 15-ton hook and ball, suspended 20 feet below the boom point on whip line
- 100-ton block reeved with four parts of line
- Lifting from the main hoist
- 125' load radius
- 345 lb of rigging

A graphic illustration of the basic configuration is shown in *Figure 41*.

Determine the gross capacity by examining the load charts. Start by reading the notes on the first page (*Figure 42A*) to ensure the configuration matches the information on the chart.

There are several points for discussion in the notes. Point 1 lists the deductions for single- and two-sheave upper boom points; they are not a part of this configuration, so no deduction is needed. Point 2 indicates that everything below the boom tip is part of the load weight, including the wire rope. Point 3 states that the crawlers must be fully extended, which matches the planned configuration. Point 4 lists the equipment and counterweight that must be installed for this chart to be valid; all the listed counterweight has been accounted for in the configuration. At Point 5, the load and whip lines are specified, along with the weight per foot; the total weight of the rope below the boom tip must be determined and included in the deductions.

The capacity chart begins with *Figure 42B*. Look for the section that applies to the #22C 230-foot boom; this is on the following page, in *Figure 42C*. Find the needed 125-foot operating radius and read across to the column for Capacity: Crawlers Extended. At 125 feet, the gross capacity is 24,100 lb. Note that the boom tip height is 202.6 feet. This information will be needed to calculate the weight of the hoist line.

RADIUS (ft)	3-SECTION BOOM OR 4-SECTION BOOM WITH 4TH RETRACTED					RADIUS (ft)	4-SECTION BOOM WITH 4TH RETRACTED			
	ON OUTRIGGERS (lb)		ON RUBBER (lb)				ON OUTRIGGERS (lb)		ON RUBBER (lb)	
	F/R	360°	F/R	360°			F/R	360°	F/R	360°
5.0	* 17,000	* 17,000	10,000	10,000	MAIN BOOM	5.0	6100	6100	6100	6100
6.0	* 15,700	* 14,400	10,000	7450		6.0	6100	6100	6100	6100
8.0	* 12,400	10,800	7150	4400		8.0	5300	5300	5300	5250
10.0	10,200	8600	4700	2950		10.0	4700	4700	4700	3750
12.0	8500	7000	3400	2175		12.0	4400	4400	3800	2450
14.0	7200	5400	2650	1700		14.0	4400	4400	2900	1875
16.0	6100	4300	2150	1325		16.0	4400	4400	2325	1525
18.0	5000	3550	1775	1100		18.0	4400	4100	1900	1225
20.0	2700	2700	1525	950		20.0	4300	3150	1600	1000
22.0	2600	2600	1350	825		22.0	3700	2700	1375	850
24.5	2200	2200	1510	675		24.5	3100	2300	1200	700
26.0	–	–	–	–	JIB	26.0	2700	2125	1010	650
28.0	–	–	–	–		28.0	2500	1900	975	575
30.0	–	–	–	–		30.0	2300	1700	850	475
31.0	–	–	–	–		31.0	2200	1625	800	425

MAIN BOOM ANGLE (deg)	JIB STRUCTURAL CAPACITIES (lb)	
	3-SECTION BOOM OR 4-SECTION BOOM WITH 4TH RETRACTED	4-SECTION BOOM WITH 4TH EXTENDED
60	3500	3500
55	3325	3325
50	3150	3150
45	3000	3000
40	2875	2875
35	2800	2800
30	2700	2700
25	2600	2600
20	2500	2500
15	2400	2400
10	2325	2325
5	2250	2250
0	2200	2200

* LIMIT RATINGS TO 11,800 LB WHEN 3RD SECTION IS EXTENDED ANY AMOUNT.

NOTES:

JIB CAPACITY IS LIMITED BY BOTH STRUCTURAL CAPACITY CHART AND MAIN CAPACITY CHART.

SHADED AREAS ARE FOVERNED BY STRUCTURAL STRENGTH, DO NOT RELY ON TIPPING.

OPERATION OF THIS EQUIPMENT IN EXCESS OF RATING CHARTS AND DISREGARD OF INSTRUCTIONS IS DANGEROUS AND VOIDS WARRANTY.

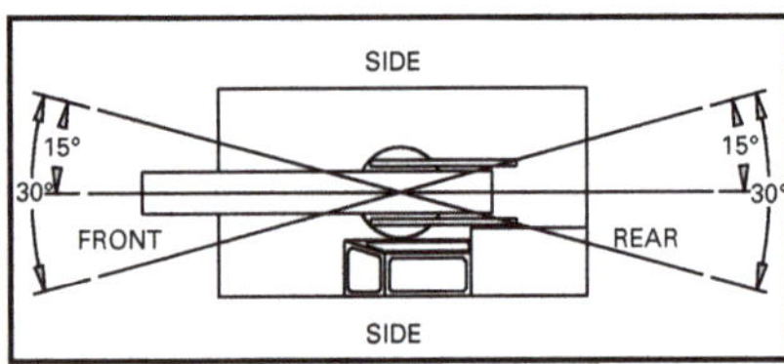

REDUCTION CHART		
	FROM MAIN BOOM RATINGS	FROM JIB RATINGS
MAIN BLOCK	140 lb	N/A
HOOK & BALL	100 lb	100 lb
JIB, STOWED	0 lb	N/A
JIB, DEPLOYED	150 lb	0 lb

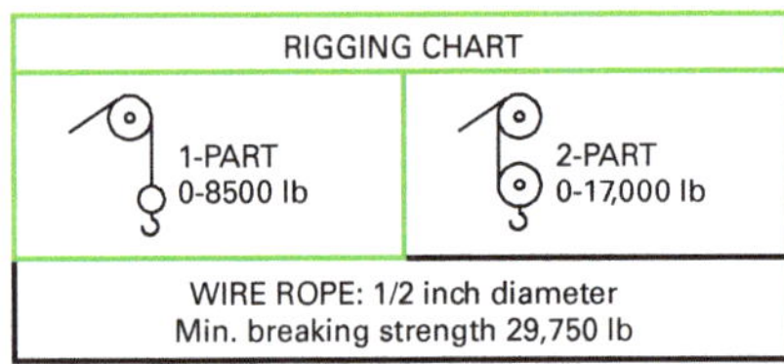

Figure 34 Grove YB4409 load chart.

1) The rated loads are the maximum lift capacities as determined by operating radius, boom extension and boom angle. The operating radius is the horizontal distance from a projection of the axis of rotation to the supporting surface, before loading, to the center of vertical hoist line or tackle with load applied.

2) The rated loads shown on outriggers do not exceed 85% of actual tipping. The rated loads shown on rubber do not exceed 75% of actual tipping. These ratings are based on freely suspended loads with the crane leveled, standing on a firm, uniform supporting surface. Practical working loads depend on supporting surface, operating radius and other factors affecting stability. Hazardous surroundings, climatic conditions, experience of personnel and proper training must all be taken into account by the operator.

3) The weights of all load handling devices such as hooks, hook blocks, slings, etc., except the hoist rope, shall be considered part of the load. See reduction chart.

4) Ratings on outriggers are for either outriggers fully extended and down or fully retracted and down. Ratings for outriggers fully retracted and down will apply for any intermediate outrigger setting.

5) Ratings on rubber depend on tire capacity, condition of tires and proper inflation pressure (100 psi). Loads on rubber may be transported at a maximum seed of 2.5 mph on a smooth, hard, level surface with boom retracted to the shortest length possible and centered over front.

6) For operating radius not shown, use load rating of next larger radius.

7) The maximum combined total boom and deck load is 12,000 lb. The maximum deck load only is 14,000 lb.

8) Do not induce any external side loads to boom or jib.

1301286YB

NCCER – *Advanced Rigger*

Figure 35 Grove TM-1500 crane.

Structural Failure Before Tipping

You might recall from an earlier section that modern cranes are more likely than older cranes to experience structural failure before tipping. The notes section of the load chart shown in *Figure 34* highlights this fact. The related note reads "Shaded areas are governed by structural strength; do not rely on tipping." This means that exceeding the crane's capacity in the shaded areas will likely result in structural failure before any sort of tipping action occurs.

NOTES FOR LIFTING CAPACITIES

WARNING: THIS CHART IS ONLY A GUIDE. The Notes below are for illustration only and should not be relied upon to operate the crane. The individual crane's load chart, operating instructions and other instruction plates must be read and understood prior to operating the crane.

1. All rated loads have been tested to and meet minimum requirements of SAE J1063 0CT80 - Cantilevered Boom Crane Structures - Method of Test, and do not exceed 85% of the tipping load on outriggers (75% of the tipping load on rubber) as determined by SAE J765 OCT 80 Crane Stability Test Code.
2. Capacities given do not include the weight of hookblocks, slings, auxiliary lifting equipment and load handling devices. Their weights MUST be added to the load to be lifted. When more than minimum required reeving is used, the additional rope weight shall be considered part of the load.
3. Capacities appearing above the bold line are based on structural strength and tipping should not be relied upon as a capacity limitation.
4. All capacities are for crane on firm, level surface. It may be necessary to have structural supports under the outrigger floats or tires to spread the load to a larger bearing surface.
5. When either boom length or radius or both are between values listed, the smallest load shown at either the next larger radius or boom length shall be used.
6. Tires shall be inflated to the recommended pressure before lifting on rubber.
7. For outrigger operation, ALL outriggers shall be fully extended with tires raised free of ground before raising the boom or lifting loads.
8. Unless otherwise stated, capacities are with powered boom sections equally extended.
9. With the tele boom extension in working position and main boom length greater than 133 ft. boom angle must not be less than 46.5°, since loss of stability will occur causing a tipping condition.

Figure 37 TM-1500 load chart notes.

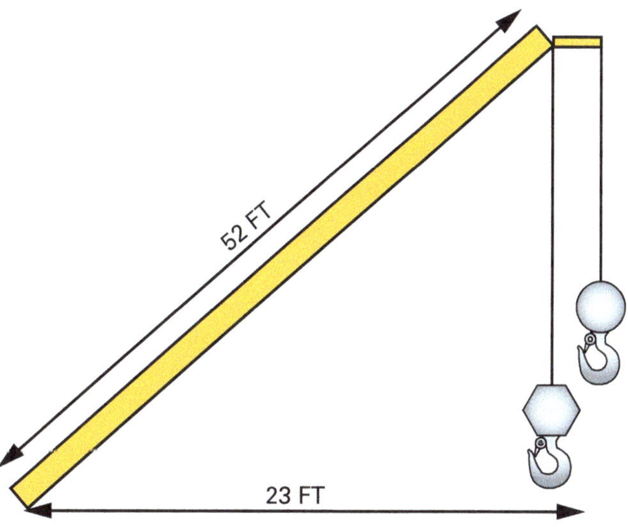

Figure 36 Graphic illustration of the TM-1500 configuration.

Rated Lifting Capacities in Pounds
46 Ft – 173 Ft Boom on Outriggers - Over Rear

| Radius in Feet | #0001 | | | | | | | | | #0002 |
| | Main Boom Length in Feet (Power Pinned Fly Retracted) | | | | | | | | | Power Pin. Fly Ext. & 141 ft. |
	46	58	70	82	94	106	118	130	141	173
10	300,000 (74.5)									
12	280,000 (72)	143,500 (76)	142,000 (79)							
15	235,000 (67.5)	143,500 (72.5)	141,500 (76.5)	130,000 (78.5)						
20	173,500 (60.5)	143,500 (67.5)	123,500 (72)	112,000 (75)	102,000 (77.5)	90,300 (79.5)				
25	136,500 (52)	131,500 (61.5)	110,500 (67.5)	98,650 (71)	89,250 (74)	78,550 (76.5)	73,700 (78.5)	69,300 (80)		
30	106,000 (43)	106,000 (55.5)	98,000 (63)	88,350 (67.5)	78,750 (71)	69,250 (73.5)	65,100 (76)	61,000 (77.5)	60,000 (79.5)	
35	84,700 (30.5)	84,700 (49)	84,700 (58)	80,150 (63.5)	69,000 (67.5)	60,750 (70.5)	57,150 (73)	54,000 (75.5)	52,150 (77.5)	
40		70,500 (41)	70,500 (52.5)	70,500 (59.5)	61,300 (64)	54,000 (67.5)	50,600 (70.5)	48,300 (73)	45,850 (75)	38,000 (79)
45		58,850 (32)	58,850 (47)	58,850 (55)	55,000 (60.5)	48,500 (64.5)	46,200 (68)	43,050 (71)	40,400 (73)	35,750 (77)
50		49,600 (17.5)	49,600 (40.5)	49,600 (50.5)	48,750 (57)	43,050 (61.5)	40,700 (65)	38,250 (68.5)	35,750 (71)	32,100 (75.5)
60			36,200 (22.5)	36,200 (39.5)	36,200 (48.5)	34,300 (55)	33,600 (59.5)	30,750 (63.5)	28,500 (66.5)	26,350 (72)
70				26,050 (25)	26,050 (39.5)	26,050 (47.5)	26,050 (53)	24,750 (58)	23,100 (61.5)	22,000 (68.5)
80					18,850 (27)	18,850 (39)	18,850 (46.5)	18,850 (52.5)	18,700 (56.5)	18,500 (64.5)
90						13,500 (28)	13,500 (38.5)	13,500 (46.5)	13,500 (51.5)	15,250 (60.5)
100							9,390 (29)	9,390 (38)	9,390 (45.5)	12,600 (56.5)
110							6,080 (12.5)	6,080 (30.5)	6,080 (39)	10,100 (52)
120								3,390 (17.5)	3,390 (31)	7,530 (47.5)
130									1,150 (19.5)	5,390 (42.5)
140										3,610 (36.5)
150										2,100 (30)
Minimum boom angle (deg.) for indicated length (no load)									10	19
Maximum boom length (ft.) at 0 deg. boom angle (no load)									140	167

Note: ( ) Boom angles are in degrees.
#LMI operating code. Refer to LMI manual for instructions.

Figure 38 Grove TM-1500 load chart.

WEIGHT REDUCTIONS FOR LOAD HANDLING DEVICES

33 ft. Extension	
*Stowed -	892 lbs.
*Erected -	5,704 lbs.

33 ft. - 58 ft. Extension	
*Stowed -	1,246 lbs.
*Erected (Ret.) -	8,412 lbs.
*Erected (Ext.) -	11,406 lbs.

46 ft. - 173 ft. Boom with	
*46 ft. Jib Erected -	9,613 lbs.
*60 ft. Jib Erected -	14,571 lbs.
*74 ft. Jib Erected -	20,443 lbs.
*88 ft. Jib Erected -	27,199 lbs.
*Fixed Jib Accessories -	327 lbs.

*Reduction of main boom capacities

HOOKBLOCKS:	
30 Ton, 1 Sheave	1,022 lbs.
150 Ton, 8 Sheave	5,254 lbs.
Auxiliary Boom Head	261 lbs.
10 Ton Headache Ball	560 lbs.
15 Ton Headache Ball	803 lbs.

Figure 39 Grove TM-1500 weight deduction table.

Now, gather the information needed for deductions. Refer to the fourth page of the load chart (*Figure 42D*). The 15-ton hook and ball requires a deduction of 865 lb. The 100-ton block requires a deduction of 2,065 lb.

The weight of the whip and load lines can be calculated using several pieces of information. Note that the weight per foot is the same for both lines. The whip line hook and ball is suspended 20 feet. With 20 feet of rope and a weight of 2.34 lb/ft, the deduction for the whip line is calculated as follows. (round the result up to the nearest whole number):

Figure 40 Manitowoc 4100W-2 lattice-boom crawler crane.

$$\begin{array}{r} 20.00 \text{ ft} \\ \times \quad 2.34 \text{ lb/ft} \\ \hline \mathbf{47.00 \text{ lb}} \end{array}$$

For the load line, it should be assumed that it will need to extend to the ground. The boom tip height at this operating radius is 202.6 feet. Recall that there are four parts of line reeved on the load block, which can make a significant difference in the figures. Calculate the deduction for the load line by first multiplying the weight (in pounds per foot) of the rope by the operating radius, as follows:

$$\begin{array}{r} 202.60 \text{ ft} \\ \times \quad 2.34 \text{ lb/ft} \\ \hline \mathbf{475.00 \text{ lb}} \end{array}$$

Multiply the result by the number of parts of line to find the total line deduction:

$$\begin{array}{r} 475 \text{ ft} \\ \times \quad 4 \text{ parts of line} \\ \hline \mathbf{1,900 \text{ lb load line deduction}} \end{array}$$

If you are using a calculator, notice that the multiplication results are rounded up to ensure that the end result remains conservative.

There is now enough information to calculate the total deductions from gross capacity, as shown below (be sure to add the weight of the rigging):

$$
\begin{aligned}
&865.00 \text{ lb (15-ton hook and ball)} \\
&2{,}065.00 \text{ lb (100-ton block)} \\
&47.00 \text{ lb (whip line)} \\
&1{,}900.00 \text{ lb (load line)} \\
+\ &345.00 \text{ lb (rigging)} \\
\hline
&\textbf{5{,}222.00 lb total deduction}
\end{aligned}
$$

The net capacity can now be calculated by subtracting the deductions from the gross capacity:

$$
\begin{aligned}
&24{,}100.00 \text{ lb gross capacity} \\
-\ &5{,}222.00 \text{ lb total deduction} \\
\hline
&\textbf{18{,}878.00 lb net capacity}
\end{aligned}
$$

3.4.4 Example Four: Boom Truck

Find the net capacity of the National Crane 18103 Series truck-mounted, telescopic-boom truck (*Figure 43*) in the following configuration:

- Fully extended 103-foot main boom, lifting from a 31-foot installed jib
- Auxiliary boom head
- 35-ton, 3-sheave block
- Outriggers fully extended
- Load radius of 68 feet
- 287 lb of rigging

A graphic illustration of this configuration is shown in *Figure 44*. In this case, the 103-foot fully extended main-boom length was taken from the range diagram in *Figure 45*. The total boom length increases to 134 feet when fully extended and the 31-foot jib is added. The appropriate arc is identified on the range diagram.

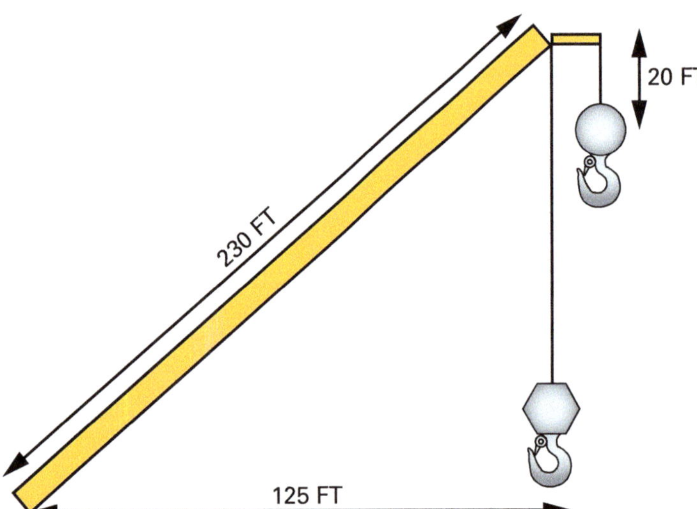

Figure 41 Graphic illustration of the Manitowoc 4100-W2 configuration.

MANITOWOC 4100 W-2

LIFTCRANE CAPACITIES

BOOM NO. 22C WITH OPEN THROAT TOP
146,400 LB. CRANE COUNTERWEIGHT
60,000 LB. CARBODY COUNTERWEIGHT
26'6" CRAWLERS EXTENDED

WARNING: This chart will apply only when two 12,000 lb. side ctwts. and two 30,000 lb. carbody ctwts. bear MEC registered Serial Numbers.

LIFTING CAPACITIES: Capacities for various boom lengths and operating radii may be based on percent of tipping, strength of structural components, operating speeds and other factors.

Capacities are for freely suspended loads and do not exceed 75% of a static tipping load. Capacities based on structural competence are shown by shaded areas.

[1] Capacities are shown in pounds. Deduct 1200 pounds from capacities listed when single sheave upper boom point is attached and 1500 pounds when two sheave upper boom point is attached. To comply with B30.5 requirements, upper boompoint cannot be used on the 260 ft. boom. [2] Weight of jib, (see chart A), all load blocks, hooks, weight ball, slings, hoist lines beneath boom and jib point sheaves, etc., is considered part of the main boom load. Boom is not to be lowered beyond radii where combined weights are greater than rated capacity. Where no capacity is shown, operation is not intended or approved.

[3] **OPERATING CONDITIONS:** Machine to operate in a level position on a firm surface with crawlers fully extended and gantry in working position and be rigged in accordance with and under conditions referred to in rigging drawing No. 190693 and load line specification chard No. 6592-A.

Crane operator judgement must be used to allow for dynamic load effects of swinging, hoisting or lowering, travel, as well as adverse operating conditions & physical machine depreciation.

OPERATOR RADIUS: Operating is the horizontal distance from the axis of rotation to the center of vertical hoist line or load block with the load freely suspended. Add 14" to boom point radius for radius of sheave when using single part hoist line.

Boom angle is the angle between horizontal and centerline of boom butt and inserts and is an indication of operating radius. In all cases, operating radius shall govern capacity.

BOOM POINT ELEVATION: Boom point elevation, in feet, is the vertical distance from ground level to centerline of boom point shaft.

MACHINE EQUIPMENT: Machine equipped with 26'6" extendible crawlers, 48" treads, 17' retractable gantry, 12 part boom hoist reeving, four 1 3/8" boom pendants, 1st ctwt. 41,900 lbs., 2nd ctwt. 41,500 lbs., 3rd ctwt. 39,000 lbs., two 12,000 lbs. side ctwt's. and two 30,000 lbs. carbody ctwt's. [4]

HOIST REEVING FOR MAIN LOAD BLOCK

No. Parts Of Line	1	2	3	4	5	6
Max Load - Lbs.	32,500	65,000	97,500	130,000	162,500	195,000
No. Parts of Line	7	8	9	10	11	12
Max. Load - Lbs.	227,500	260,000	292,500	325,000	357,500	400,000
No. Parts of Line	13					
Max. Load - Lbs.	430,000					

LOAD AND WHIP LINE SPECIFICATIONS [5]

LOAD LINE:	1-1/8" - 6 x 31 Warrington-Seale, Extra Improved Plow Steel, Regular Lay, IWRC. Minimum Breaking Strength 65 Ton. (Approx. Weight Per Ft. in Lbs. 2.34)
WHIP LINE:	1-1/8" - Warrington-Seale, Improved Plow Steel, Regular Lay, IWRC. Minimum Breaking Strength 56.5 Ton. Maximum Load - 28,300 Lbs. Per Line. (Approx. Weight Per Ft. in Lbs. 2.34)

MAXIMUM BOOM AND JIB LENGTH LIFTED UNASSISTED				DEDUCT FROM CAPACITIES WHEN JIB IS ATTACHED	
OVER FRONT OF BLOCKED CRAWLERS		OVER SIDE OF EXTENDED CRAWLERS			
BOOM LENGTH	JIB NO. 123	BOOM LENGTH	JIB NO. 123	JIB LENGTH	JIB NO. 123
260'	--	260'	--	30'	3,000 lbs.
250'	--	250'	--	40'	3,600 lbs.
240'	40'	240'	40'	50'	4,200 lbs.
230'	60'	230'	60'	60'	4,900 lbs.

Load block, hook and weight ball on ground to start.

FOR JIB CAPACITIES, CONSULT JIB CHART.

BOOM LGTH FEET	OPER. RAD. FEET	BOOM ANG. DEG.	BOOM POINT ELEV.	CAPACITY: CRAWLERS EXTENDED
70	16.5	79.7	75.9	460,000
	17	79.3	75.8	400,000
	18	78.5	75.6	380,100
	19	77.6	75.4	363,000
	20	76.8	75.1	347,300
	22	75.1	74.6	319,600
	24	73.4	74.1	293,400
	26	71.7	73.5	266,100
	28	69.9	72.8	237,500
	30	68.2	72.0	214,300
	32	66.4	71.2	195,100
	34	64.6	70.2	178,900
	36	62.8	69.3	165,200
	38	60.9	68.2	153,300
	40	59.1	67.0	143,000
	45	54.1	63.7	122,100
	50	48.9	59.8	106,300
	55	43.2	54.9	93,900
	60	36.9	49.0	84,000
	65	29.4	41.3	75,800
	70	19.5	30.3	63,900

BOOM LGTH FEET	OPER. RAD. FEET	BOOM ANG. DEG.	BOOM POINT ELEV.	CAPACITY: CRAWLERS EXTENDED
80	17	80.6	85.9	392,800
	18	79.9	85.8	378,900
	19	79.2	85.6	361,800
	20	78.5	85.4	346,100
	22	77.0	84.9	318,400
	24	75.5	84.5	292,500
	26	74.0	83.9	265,600
	28	72.5	83.3	237,000
	30	71.0	82.7	213,800
	32	69.5	81.9	194,600
	34	68.0	81.2	178,500
	36	66.4	80.3	164,700
	38	64.8	79.4	152,800
	40	63.3	78.4	142,400
	45	59.2	75.7	121,500
	50	54.9	72.5	105,700
	55	50.4	68.6	93,300
	60	45.6	64.1	83,400
	65	40.3	58.8	75,200
	70	34.4	52.2	68,300
	75	27.4	43.9	52,500
	80	18.2	32.0	53,900

BOOM LGTH FEET	OPER. RAD. FEET	BOOM ANG. DEG.	BOOM POINT ELEV.	CAPACITY: CRAWLERS EXTENDED
90	18	81.1	95.9	355,400
	19	80.4	95.7	346,900
	20	79.8	95.6	336,900
	22	78.5	95.2	317,400
	24	77.2	94.7	291,700
	26	75.9	94.3	264,800
	28	74.5	93.7	236,600
	30	73.2	93.2	213,400
	32	71.9	92.5	194,200
	34	70.5	91.9	178,000
	36	69.2	91.1	164,200
	38	67.8	90.3	152,300
	40	66.4	89.5	142,000
	45	62.9	87.1	121,100
	50	59.3	84.4	105,200
	55	55.5	81.2	92,800
	60	51.5	77.5	82,900
	65	47.3	73.2	74,700
	70	42.8	68.2	67,800
	75	37.9	62.3	62,000
	80	32.4	55.2	57,000
	85	25.8	46.2	52,600
	90	17.1	33.5	45,900

BOOM LGTH FEET	OPER. RAD. FEET	BOOM ANG. DEG.	BOOM POINT ELEV.	CAPACITY: CRAWLERS EXTENDED
100	19	81.4	105.9	332,900
	20	80.8	105.7	327,100
	22	79.6	105.4	316,200
	24	78.5	105.0	290,800
	26	77.3	104.5	263,900
	28	76.1	104.1	236,200
	30	74.9	103.6	212,900
	32	73.7	103.0	193,700
	34	72.5	102.4	177,500
	36	71.3	101.7	163,700
	38	70.1	101.0	151,800
	40	68.9	100.3	141,400
	45	65.8	98.2	120,500
	50	62.6	95.8	104,700
	55	59.3	93.0	92,300
	60	55.9	89.8	82,300
	65	52.4	86.2	74,100
	70	48.7	82.1	67,200
	75	44.8	77.4	61,400
	80	40.5	72.0	56,400
	85	35.9	65.6	52,000
	90	30.7	58.0	48,200
	95	24.5	48.5	44,900
	100	16.3	35.0	39,300

CAUTION! CHECK AMOUNT OF COUNTERWEIGHT ON MACHINE BEFORE USE OF THIS CHART

***Figure* 42A** Manitowoc 4100 W-2 load chart (1 of 4).

BOOM LGTH FEET	OPER. RAD. FEET	BOOM ANG. DEG.	BOOM POINT ELEV.	CAPACITY: CRAWLERS EXTENDED
110	22	82.3	115.7	240,500
	24	81.3	115.4	232,100
	26	80.2	115.0	224,400
	28	79.1	114.6	217,300
	30	78.1	114.1	210,700
	32	77.0	113.6	193,300
	34	75.9	113.1	177,100
	36	74.8	112.5	163,300
	38	73.7	111.8	151,400
	40	72.6	111.2	141,000
	45	69.9	109.3	120,100
	50	67.0	107.2	104,200
	55	64.1	104.7	91,800
	60	61.2	101.9	81,800
	65	58.1	98.8	73,600
	70	54.9	95.3	66,800
	75	51.6	91.3	60,900
	80	48.1	86.8	55,900
	85	44.4	81.7	51,600
	90	40.4	75.9	47,800
	95	36.1	69.2	44,400
	100	31.1	61.1	41,400
	105	25.3	51.1	38,700
	110	17.6	37.2	33,900
120	22	83.0	125.9	228,700
	24	82.0	125.5	220,500
	26	81.0	125.2	213,000
	28	80.0	124.8	206,100
	30	79.1	124.4	199,700
	32	78.1	123.9	192,800
	34	77.1	123.4	176,600
	36	76.1	122.9	162,800
	38	75.1	122.3	150,900
	40	74.1	121.7	140,500
	45	71.6	120.0	119,500
	50	69.1	118.1	103,700
	55	66.5	115.9	91,300
	60	63.8	113.4	81,300
	65	61.1	110.6	73,100
	70	58.3	107.5	66,200
	75	55.4	104.0	60,400
	80	52.4	100.1	55,300
	85	49.2	95.8	51,000
	90	45.9	91.0	47,200
	95	42.4	85.5	43,800
	100	38.6	79.4	40,800
	105	34.4	72.2	38,100
	110	29.7	63.7	35,700
	115	24.1	53.2	33,500
130	24	82.6	135.7	212,700
	26	81.7	135.3	205,300
	28	80.8	135.0	198,600
	30	79.9	134.6	192,300
	32	79.0	134.2	186,500
	34	78.1	133.7	176,300
	36	77.2	133.2	162,500
	38	76.3	132.7	150,600
	40	75.4	132.1	140,200
	45	73.1	130.6	119,300
	50	70.8	128.8	103,400
	55	68.4	126.8	91,000
	60	66.0	124.6	81,000
	65	63.5	122.0	72,800
	70	61.0	119.2	65,900
	75	58.4	116.1	60,000
	80	55.7	112.7	55,000
	85	53.0	108.9	50,700
	90	50.1	104.8	46,800
	95	47.1	100.1	43,500
	100	43.9	95.0	40,500
	105	40.6	89.2	37,800
	110	36.9	82.6	35,300
	115	33.0	75.1	33,100
	120	28.5	66.1	31,100
	125	23.1	55.1	29,300
140	26	82.3	145.5	198,400
	28	81.5	145.1	191,800
	30	80.7	144.8	185,600
	32	79.8	144.4	179,900
	34	79.0	144.0	174,700
	36	78.1	143.5	162,000
	38	77.3	143.0	150,100
	40	76.5	142.5	139,700
	45	74.3	141.1	118,700
	50	72.2	139.4	102,800
	55	70.0	137.6	90,400
	60	67.8	135.5	80,400
	65	65.6	133.2	72,200
	70	63.3	130.7	65,300
	75	60.9	127.9	59,500
	80	58.5	124.8	54,400
	85	56.1	121.4	50,100
	90	53.5	117.7	46,200
	95	50.9	113.7	42,900
	100	48.1	109.2	39,900

BOOM LGTH FEET	OPER. RAD. FEET	BOOM ANG. DEG.	BOOM POINT ELEV.	CAPACITY: CRAWLERS EXTENDED
140	105	45.3	104.2	37,200
	110	42.2	98.8	34,700
	115	39.0	92.7	32,500
	120	35.5	85.7	30,500
	125	31.7	77.8	28,700
	130	27.4	68.5	27,000
	135	22.2	57.0	25,500
150	28	82.1	155.3	185,600
	30	81.3	154.9	179,600
	32	80.5	154.6	174,000
	34	79.7	154.2	168,800
	36	79.0	153.8	161,600
	38	78.2	153.3	149,600
	40	77.4	152.8	139,200
	45	75.4	151.5	118,200
	50	73.4	150.0	102,400
	55	71.4	148.3	89,900
	60	69.4	146.4	79,900
	65	67.3	144.3	71,700
	70	65.2	141.9	64,800
	75	63.0	139.4	59,000
	80	60.9	136.6	53,900
	85	58.6	133.5	49,600
	90	56.3	130.1	45,700
	95	54.0	126.5	42,400
	100	51.5	122.5	39,400
	105	49.0	118.2	36,700
	110	46.4	113.4	34,200
	115	43.6	108.2	32,000
	120	40.7	102.4	30,000
	125	37.6	96.0	28,200
	130	34.2	88.8	26,500
	135	30.5	80.5	25,000
	140	26.4	70.8	23,500
	145	21.4	58.8	22,200
160	28	82.6	165.4	179,800
	30	81.8	165.1	173,900
	32	81.1	164.7	168,400
	34	80.4	164.4	163,300
	36	79.7	164.0	158,600
	38	78.9	163.6	149,100
	40	78.2	163.1	138,700
	45	76.3	161.9	117,700
	50	74.5	160.5	101,800
	55	72.6	158.9	89,300
	60	70.7	157.1	79,300
	65	68.8	155.1	71,100
	70	66.8	153.0	64,200
	75	64.9	150.6	58,400
	80	62.9	148.0	53,300
	85	60.8	145.2	49,000
	90	58.7	142.2	45,100
	95	56.6	138.8	41,800
	100	54.4	135.2	38,700
	105	52.1	131.4	36,000
	110	49.7	127.1	33,600
	115	47.3	122.5	31,400
	120	44.8	117.5	29,400
	125	42.1	112.0	27,600
	130	39.3	105.9	25,900
	135	36.3	99.2	24,300
	140	33.1	91.7	22,900
	145	29.5	83.0	21,600
	150	25.5	72.9	20,400
	155	20.7	60.5	19,200
170	30	82.3	175.2	169,000
	32	81.6	174.9	163,600
	34	81.0	174.5	158,700
	36	80.3	174.2	154,000
	38	79.6	173.8	148,800
	40	78.9	173.3	138,400
	45	77.2	172.2	117,400
	50	75.4	170.9	101,500
	55	73.7	169.4	89,100
	60	71.9	167.7	79,000
	65	70.1	165.9	70,800
	70	68.3	163.9	63,900
	75	66.4	161.7	58,100
	80	64.6	159.3	53,000
	85	62.7	156.7	48,700
	90	60.8	153.9	44,900
	95	58.8	150.8	41,500
	100	56.8	147.5	38,500
	105	54.7	144.0	35,800
	110	52.6	140.2	33,300
	115	50.4	136.0	31,100
	120	48.1	131.5	29,100
	125	45.8	126.7	27,300
	130	43.3	121.4	25,600
	135	40.8	115.7	24,100
	140	38.1	109.3	22,600
	145	35.2	102.3	21,300
	150	32.0	94.5	20,100
	155	28.6	85.5	18,900
	160	24.7	75.1	17,800
	165	20.0	62.2	16,400

BOOM LGTH FEET	OPER. RAD. FEET	BOOM ANG. DEG.	BOOM POINT ELEV.	CAPACITY: CRAWLERS EXTENDED
180	32	82.1	185.0	158,700
	34	81.5	184.7	154,000
	36	80.8	184.3	149,500
	38	80.2	184.0	145,200
	40	79.5	183.6	137,900
	45	77.9	182.5	116,900
	50	76.3	181.2	101,000
	55	74.6	179.8	88,500
	60	72.9	178.3	78,500
	65	71.3	176.6	70,300
	70	69.6	174.7	63,400
	75	67.8	172.6	57,500
	80	66.1	170.4	52,500
	85	64.3	168.0	48,100
	90	62.5	165.4	44,300
	95	60.7	162.5	40,900
	100	58.9	159.5	37,900
	105	57.0	156.2	35,200
	110	55.0	152.7	32,700
	115	53.0	149.0	30,500
	120	51.0	144.9	28,500
	125	48.9	140.5	26,700
	130	46.7	135.8	25,000
	135	44.4	130.7	23,500
	140	42.0	125.2	22,000
	145	39.6	119.2	20,700
	150	36.9	112.6	19,500
	155	34.1	105.3	18,300
	160	31.1	97.2	17,200
	165	27.7	87.9	16,200
	170	23.9	77.1	15,300
	175	19.4	63.8	13,500
190	32	82.5	195.1	153,100
	34	81.9	194.8	148,800
	36	81.3	194.5	144,800
	38	80.7	194.1	140,800
	40	80.1	193.8	137,000
	45	78.5	192.7	116,400
	50	77.0	191.5	100,500
	55	75.4	190.2	88,000
	60	73.9	188.8	78,000
	65	72.3	187.1	69,700
	70	70.7	185.4	62,800
	75	69.1	183.4	57,000
	80	67.4	181.4	52,000
	85	65.8	179.1	47,600
	90	64.1	176.6	43,800
	95	62.4	174.0	40,400
	100	60.7	171.2	37,400
	105	58.9	168.2	34,700
	110	57.1	164.9	32,200
	115	55.3	161.4	30,000
	120	53.4	157.7	28,000
	125	51.5	153.7	26,200
	130	49.5	149.5	24,500
	135	47.5	144.9	22,900
	140	45.3	140.0	21,500
	145	43.1	134.6	20,200
	150	40.8	128.9	18,900
	155	38.4	122.6	17,800
	160	35.9	115.8	16,700
	165	33.2	108.3	15,700
	170	30.2	99.8	14,800
	175	27.0	90.3	13,900
	180	23.3	79.1	13,100
	185	18.9	65.4	11,000
200	34	82.3	204.9	143,100
	36	81.7	204.6	139,500
	38	81.2	204.3	136,000
	40	80.6	203.9	132,500
	45	79.1	203.0	115,900
	50	77.7	201.8	100,000
	55	76.2	200.6	87,500
	60	74.7	199.2	77,400
	65	73.2	197.7	69,200
	70	71.7	196.0	62,300
	75	70.2	194.2	56,400
	80	68.6	192.2	51,400
	85	67.1	190.1	47,000
	90	65.5	187.8	43,200
	95	63.9	185.3	39,800
	100	62.3	182.7	36,800
	105	60.6	179.8	34,100
	110	59.0	176.8	31,600
	115	57.3	173.6	29,400
	120	55.5	170.2	27,400
	125	53.0	166.5	25,600
	130	51.9	162.6	23,900
	135	50.1	158.4	22,300
	140	48.1	153.9	20,900
	145	46.2	149.1	19,600
	150	44.1	144.0	18,300
	155	42.0	138.4	17,200
	160	39.7	132.5	16,100
	165	37.4	126.0	15,100
	170	34.9	118.9	14,200

BOOM LGTH FEET	OPER. RAD. FEET	BOOM ANG. DEG.	BOOM POINT ELEV.	CAPACITY: CRAWLERS EXTENDED
200	175	32.3	111.1	13,300
	180	29.4	102.4	12,400
	185	26.2	92.5	11,700
	190	22.6	81.0	10,900
	195	18.4	67.0	8,500
210	36	82.1	214.7	134,200
	38	81.6	214.4	131,000
	40	81.0	214.1	127,900
	45	79.6	213.2	115,500
	50	78.3	212.1	99,600
	55	76.9	210.9	87,100
	60	75.4	209.6	77,100
	65	74.0	208.1	68,800
	70	72.6	206.6	61,900
	75	71.2	204.8	56,100
	80	69.7	203.0	51,000
	85	68.2	201.0	46,700
	90	66.7	198.8	42,800
	95	65.2	196.5	39,400
	100	63.7	194.0	36,400
	105	62.2	191.3	33,700
	110	60.6	188.5	31,300
	115	59.0	185.5	29,100
	120	57.4	182.3	27,100
	125	55.7	178.9	25,200
	130	54.1	175.2	23,500
	135	52.3	171.4	22,000
	140	50.6	167.3	20,600
	145	48.8	162.9	19,200
	150	46.9	158.2	18,000
	155	45.0	153.2	16,800
	160	43.0	147.9	15,800
	165	40.9	142.1	14,800
	170	38.7	135.9	13,800
	175	36.5	129.2	12,900
	180	34.0	121.9	12,100
	185	31.5	113.9	11,300
	190	28.7	104.9	10,600
	195	25.6	94.7	9,900
	200	22.1	82.9	8,800
	205	17.9	68.5	6,500
220	38	82.0	224.6	125,700
	40	81.4	224.2	122,800
	45	80.1	223.3	115,000
	50	78.8	222.3	99,100
	55	77.5	221.2	86,600
	60	76.1	220.0	76,600
	65	74.8	218.6	68,300
	70	73.4	217.1	61,400
	75	72.0	215.4	55,500
	80	70.7	213.7	50,500
	85	69.3	211.8	46,100
	90	67.9	209.7	42,300
	95	66.4	207.5	38,900
	100	65.0	205.2	35,800
	105	63.6	202.7	33,100
	110	62.1	200.0	30,700
	115	60.6	197.2	28,500
	120	59.1	194.2	26,500
	125	57.5	191.0	24,600
	130	56.0	187.6	23,000
	135	54.3	184.0	21,400
	140	52.7	180.2	20,000
	145	51.0	176.1	18,600
	150	49.3	171.8	17,400
	155	47.6	167.3	16,300
	160	45.7	162.4	15,200
	165	43.9	157.2	14,200
	170	41.9	151.7	13,200
	175	39.9	145.7	12,300
	180	37.8	139.3	11,500
	185	35.6	132.4	10,700
	190	33.2	124.8	10,000
	195	30.7	116.6	9,300
	200	28.0	107.3	8,600
	205	25.0	96.9	8,000
	210	21.5	84.8	6,500

CAPACITIES CONTINUED ON NEXT PAGE

These load charts are intended for instructional purposes only. They were derived from manufacturer sales information which may not be complete or machine specific. Not responsible for typographical errors.

Figure 42B Manitowoc 4100 W-2 load chart (2 of 4).

SEE CONDITIONS ON FRONT PAGE

BOOM LGTH FEET	OPER. RAD. FEET	BOOM ANG. DEG.	BOOM POINT ELEV.	CAPACITY: CRAWLERS EXTENDED
2 3 0	38	81.5	234.5	132,900
	40	81.0	234.2	125,200
	45	79.7	233.3	114,500
	50	78.5	232.4	98,600
	55	77.2	231.3	86,100
	60	75.9	230.1	76,100
	65	74.6	228.8	67,800
	70	73.3	227.3	60,900
	75	72.0	225.8	55,000
	80	70.7	224.1	50,000
	85	69.4	222.3	45,600
	90	68.0	220.3	41,700
	95	66.7	218.2	38,400
	100	65.3	216.0	35,300
	105	64.0	213.6	32,600
	110	62.6	211.1	30,200
	115	61.1	208.4	28,000
	120	59.7	205.6	26,000
	125	58.3	202.6	24,100
	130	56.8	199.4	22,400
	135	55.3	196.0	20,900
	140	53.8	192.5	19,400
	145	52.2	188.7	18,100
	150	50.6	184.7	16,900
	155	49.0	180.5	15,700
	160	47.3	176.0	14,700
	165	45.6	171.3	13,600
	170	43.8	166.2	12,700
	175	42.0	160.8	11,800
	180	40.1	155.1	11,000
	185	38.1	148.9	10,200
	190	36.0	142.3	9,400
	195	33.9	135.1	8,700
	200	31.6	127.3	8,100
	205	29.1	118.8	7,400
	210	26.4	109.3	6,800
2 4 0	40	81.4	244.3	123,400
	45	80.2	243.5	112,600
	50	78.9	242.6	98,100
	55	77.7	241.5	85,600
	60	76.5	240.4	75,600
	65	75.3	239.1	67,300
	70	74.0	237.7	60,400
	75	72.8	236.3	54,500
	80	71.5	234.6	49,500
	85	70.3	232.9	45,100
	90	69.0	231.1	41,200
	95	67.7	229.1	37,800
	100	66.4	227.0	34,800
	105	65.1	224.7	32,100
	110	63.8	222.3	29,700
	115	62.5	219.8	27,400
	120	61.1	217.1	25,400
	125	59.7	214.3	23,600
	130	58.3	211.3	21,900
	135	56.9	208.1	20,300
	140	55.5	204.7	18,900
	145	54.0	201.2	17,600
	150	52.5	197.5	16,300
	155	51.0	193.5	15,200
	160	49.5	189.4	14,100
	165	47.9	185.0	13,100
	170	46.2	180.3	12,200
	175	44.6	175.4	11,300
	180	42.8	170.2	10,400
	185	41.0	164.6	9,600
	190	39.2	158.7	8,900
	195	37.3	152.3	8,200
	200	35.2	145.5	7,500
	205	33.1	138.1	6,700
	210	30.9	130.1	6,000

BOOM LGTH FEET	OPER. RAD. FEET	BOOM ANG. DEG.	BOOM POINT ELEV.	CAPACITY: CRAWLERS EXTENDED
2 5 0	45	80.6	253.6	106,600
	50	79.4	252.7	97,800
	55	78.2	251.7	85,200
	60	77.1	250.6	75,200
	65	75.9	249.4	66,900
	70	74.7	248.1	60,000
	75	73.5	246.7	54,100
	80	72.3	245.2	49,100
	85	71.1	243.5	44,700
	90	69.9	241.7	40,800
	95	68.7	239.8	37,400
	100	67.4	237.8	34,400
	105	66.2	235.7	31,700
	110	64.9	233.4	29,300
	115	63.6	231.0	27,100
	120	62.4	228.5	25,000
	125	61.1	225.8	23,200
	130	59.7	222.9	21,500
	135	58.4	219.9	20,000
	140	57.0	216.8	18,500
	145	55.7	213.4	17,200
	150	54.3	209.9	16,000
	155	52.8	206.2	14,800
	160	51.4	202.4	13,700
	165	49.9	198.3	12,700
	170	48.4	193.9	11,800
	175	46.8	189.4	10,900
	180	45.3	184.5	10,100
	185	43.6	179.4	9,300
	190	41.9	174.0	8,500
	195	40.2	168.3	7,800
	200	38.4	162.2	7,000
	205	36.5	155.7	6,200
	210	34.5	148.6	5,500
2 6 0	45	80.9	263.7	104,800
	50	79.8	262.9	95,700
	55	78.7	261.9	84,700
	60	77.6	260.9	74,700
	65	76.4	259.7	66,400
	70	75.3	258.5	59,500
	75	74.2	257.1	53,600
	80	73.0	255.6	48,500
	85	71.8	254.1	44,100
	90	70.7	252.4	40,300
	95	69.5	250.6	36,900
	100	68.3	248.6	33,900
	105	67.1	246.6	31,100
	110	65.9	244.4	28,700
	115	64.7	242.1	26,500
	120	63.5	239.7	24,500
	125	62.3	237.1	22,600
	130	61.0	234.4	20,900
	135	59.7	231.6	19,400
	140	58.5	228.6	17,900
	145	57.2	225.4	16,600
	150	55.8	222.1	15,400
	155	54.5	218.7	14,200
	160	53.1	215.0	13,100
	165	51.7	211.2	12,100
	170	50.3	207.1	11,200
	175	48.9	202.9	10,300
	180	47.4	198.4	9,500
	185	45.9	193.7	8,700
	190	44.3	188.7	7,900
	195	42.7	183.4	7,000
	200	41.1	177.8	6,200
	205	39.4	171.9	5,400
	210	37.6	165.6	4,700

© Manitowoc 2017

Figure 42C Manitowoc 4100 W-2 load chart (3 of 4).

DESCRIPTION	APPROX. WEIGHT (IN LBS.)

JIB NO. 123

Jib Top - 15' (w/Jib point)	695
Jib Butt - 15'	690
Jib Insert - 10'	340
Basic Pendant - 33' 3-3/4" (2 Req'd)	115 each
Pendant - 10' (2 Per insert)	65 each
Jib Backstay Pendant	155 each
Jib Strut - 12'- 6"	365

COMPONENTS

Hook Rollers (6) - w/shafts	1,020
Light Plant - 6.5KW - w/mounting platform	1,390
Catwalk - left and right side w/rails	1,320
Lagging - 27-5/8" dia. plain	1,410
Boom Hoist Rope - 12 Part - 760' of 7/8" - 6 x 26	1,080
Wire Rope Guide Assembly - Lower	325
Wire Rope Guide Assembly - Upper	510
Rope Guide Roller Assembly	55 each
2-Part Gantry w/Telescopic Backhitch	7,805
Equalizer	2,000
Hoist Line - 1-1/8" - 6 x 31	2.34 lbs./ft.
Whip Line - 1-1/8" - 6 x 31	2.34 lbs./ft.
15-Ton Hook and Weight Ball	865
100-Ton Hook Block Assembly	2,065
200-Ton Hook Block Assembly	4,900
230-Ton Hook Block Assembly	5,375
Boom Stop - Telescopic Air Cushioned	675
Dragline Fairlead - Revolving	1,910
Dragline Fairlead - Hinged	9,330

NOTE: The above weights may fluctuate up or down 5% due to manufacturing tolerances.

Figure 42D Manitowoc 4100 W-2 load chart (4 of 4).

Figure 43 NC 1800 Series boom truck.

Always begin by determining the gross capacity. The best practice is to review the load chart notes first. The notes in *Figure 45* indicate that any accessories attached to the boom or load line must be deducted from the load chart capacities. The notes also indicate that when operating with the jib, the operator must refer to the jib-rated load sections of the chart. That section of the load chart has been highlighted.

The chart doesn't have an entry for a 68-foot load radius, so select the next higher radius of 70 feet. The gross capacity shown for a 70-foot load radius is 4,100 lb.

Next, identify all the applicable deductions. The deduction table shown in *Figure 46* indicates a deduction of 100 lb for the auxiliary boom head, and 870 lb for the 35-ton load block on the crane.

The required deductions are added together, as shown here (again, include the weight of the rigging):

$$
\begin{array}{r}
100.00 \text{ lb (auxiliary boom head)} \\
870.00 \text{ lb (3-sheave block)} \\
+ \quad 287.00 \text{ lb (rigging)} \\
\hline
\textbf{1,257.00 lb total deduction}
\end{array}
$$

The final step of determining the net capacity can now be completed:

$$
\begin{array}{r}
4,100.00 \text{ lb gross capacity} \\
- \quad 1,257.00 \text{ lb total deduction} \\
\hline
\textbf{2,843.00 lb net capacity}
\end{array}
$$

These have been basic examples of different crane types and configurations. Operators will certainly encounter more complex scenarios. While it is not unusual to find load charts a little confusing at first, remember that most crane operators aren't switching between drastically different cranes on a daily basis. In many cases, an operator works with the same crane, or a similar one, for days and perhaps weeks and months.

As an operator, you will become familiar with your crane's documentation and will soon be able to navigate the charts quickly and accurately. You may even begin to memorize load chart information and gross capacities. However, it is never a good idea to rely on your memory when it comes to load charts. Always take the time to review the load chart carefully, since errors can be catastrophic.

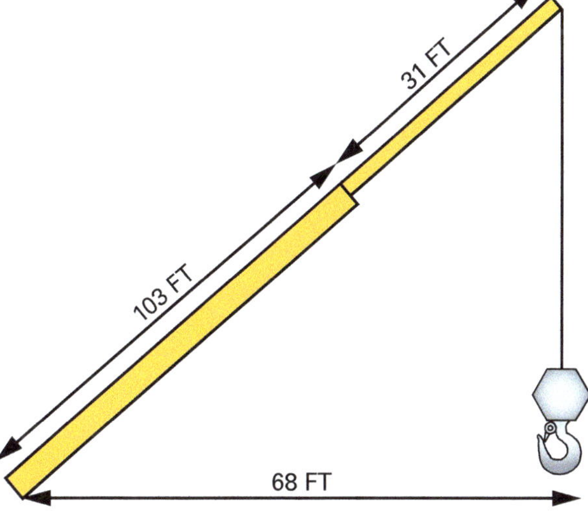

Figure 44 Graphic illustration of the National 18103 boom truck configuration.

National Crane will send you a chart on request – or you may secure needed load rating information through your nearest National Crane dealer.

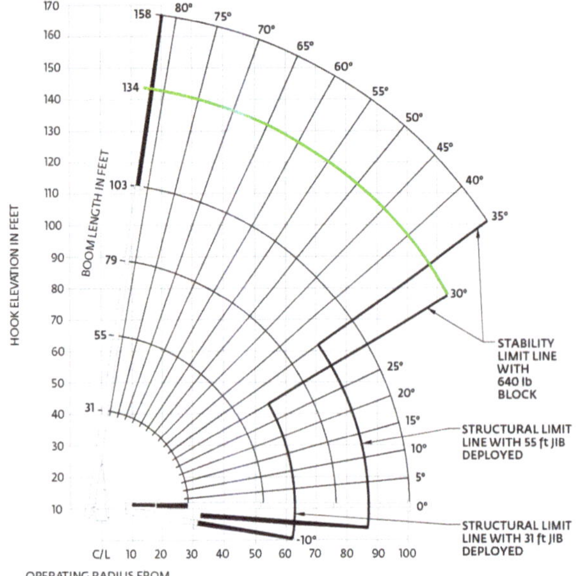

CAUTION:
- **Do not operate crane booms, jib extensions, any accessories or loads within 3 m (10 ft) of live power lines or other conductors of electricity.**
- Jib and boom capacities shown are maximum for each section.
- Do not exceed capacities at reduced radii.
- Load ratings shown on the load rating charts are maximum allowable loads with the outriggers properly extended on a firm, level surface and the crane leveled and mounted on a factory recommended truck.
- Always level the crane with the level indicator located on the crane.
- The operator must reduce load to allow for factors such as wind, ground conditions, operating speeds and their effects on freely suspended loads.
- Overloading this crane may cause structural collapse or instability.
- Weights on any accessories attached to the boom or loadline must be deducted from the load chart capacities.
- Do not exceed jib capabilities at any reduced boom lengths.
- Do not deadhead lineblock against boom tip when extending boom or winching up.
- Keep at least three wraps of loadline on drum at all times.
- Use only specified cable with this machine.

Load chart

31 ft - 103 ft BOOM RATED LOADS WITHOUT JIB

LOAD RADIUS (ft)	LOADED BOOM ANGLE	31 FT BOOM (lb)	LOADED BOOM ANGLE	55 FT BOOM (lb)	LOADED BOOM ANGLE	79 FT BOOM (lb)	LOADED BOOM ANGLE	103 FT BOOM (lb)
7	73.9	80,000						
8	71.9	74,000						
10	67.7	65,000	78.9	50,000				
12	63.5	57,000	76.6	45,000				
15	56.7	44,000	73.3	38,000	79.6	30,000		
20	44.1	30,800	67.7	31,500	75.9	26,000	79.5	17,000
25	27.4	23,200	61.7	23,800	72.1	22,000	76.7	15,200
30			55.3	18,800	68.1	18,500	73.8	13,500
35			48.3	15,200	64	15,500	70.9	12,000
40			40.5	12,500	59.6	12,800	67.8	10,500
45			31.2	10,500	55.1	10,700	65	9300
50			19.3	9000	50.7	9000	61.8	8300
55					45.5	7600	58.5	7400
60					39.9	6600	55.1	6500
65					33.4	5600	51.4	5600
70					25.5	4800	47.5	4800
75					13.4	4050	43.4	4100
80							38.9	3500
85							33.8	2950
90							28	2450
95							20.7	2050
100							7.9	1650
	0	19,700	0	8200	0	3800	0	1600

31 ft JIB RATED LOADS

RADIUS FULLY EXTENDED	LOADED BOOM ANGLE	RATED LOADS ALL BOOM LENGTHS
25	80	8800
38	75	8000
49	70	6500
60	65	5100
70	60	4100
79	55	3300
88	50	2600
96	45	1900
103	40	1350
110	35	950
115	30	650

55 ft JIB RATED LOADS

RADIUS FULLY EXTENDED	LOADED BOOM ANGLE	RATED LOADS ALL BOOM LENGTHS
29	80	4000
45	75	3700
59	70	3300
73	65	3000
85	60	2600
96	55	2100
106	50	1700
115	45	1300
123	40	950
130	35	650

NOTE:
1. Operate with jib by radius when main boom is fully extended. If necessary increase boom angle to maintain loaded radius.
2. Operate with jib by boom angle when main boom is not fully extended. Do not exceed rated jib capacities at any reduced boom lengths.
3. Capacities do not exceed 85% stability.
4. Shaded areas are structurally limited capacities.

NOTE:
1. All capacities are in pounds, angles in degrees, radius in feet.
2. Loaded boom angles are given as reference only.
3. Shaded areas are structurally limited capacities.

RATED LOAD REDUCTIONS WITH JIB

BOOM LENGTH	31 ft-55 ft JIB STOWED	31 ft-55 ft JIB ERECTED AT 31 ft LENGTH
31 ft	Reduce load 800 lb	Reduce load 2300 lb
55 ft	Reduce load 450 lb	Reduce load 2000 lb
79 ft	Reduce load 350 lb	Reduce load 1900 lb
103 ft	Reduce load 250 lb	Reduce load 1800 lb

Figure 45 NC 1800 Series load chart.

Loadline deduct		
	Aux boom head	45 kg (100 lb)
5 USt	Downhaul weight	82 kg (180 lb)
15 USt	1-sheave block	170 kg (375 lb)
25 USt	2-sheave block	290 kg (640 lb)
35 USt	3-sheave block	395 kg (870 lb)
40 USt	4-sheave block	440 kg (970 lb)

Figure 46 NC 1800 Series boom truck deductions.

Additional Resources

ASME Standard B30.5, Mobile and Locomotive Cranes. Current edition. New York, NY: American Society of Mechanical Engineers.

Cranes: Design, Practice, and Maintenance, Ing J. Verschoof. 2002. Hoboken, NJ: John Wiley and Sons, Inc.

29 *CFR* 1926, Subpart CC, *Cranes and Derricks in Construction.* **www.ecfr.gov**

3.0.0 Section Review

1. Load chart capacity deductions are usually required for all of the following *except* _____.

   a. a jib extension when hoisting on the main boom
   b. the main boom as built by the manufacturer
   c. multi-sheave blocks and headache balls
   d. an auxiliary boom head

2. Rotation-resistant wire rope is made by _____.

   a. using fiber-core wire rope
   b. using synthetic rope for the intermediate strands
   c. applying lubricants after each strand is applied
   d. applying strand layers in opposing directions

3. At what point in a lift operation is the best time to consult the crane's range diagram?

   a. Just prior to commencing the lift
   b. At the start of lift planning before any capacity determinations are done
   c. Only if a problem arises during the lift that requires adjusting the crane configuration
   d. Not required if the crane has already been erected

4. Refer to the YB4409 load chart shown in *Figure 34*. Which of the following is a *true* statement?

   a. When the jib is extended, the maximum load radius is 26 feet, regardless of load weight.
   b. At a 16-foot load radius on rubber, the crane has more gross capacity when the fourth section of the main boom is extended.
   c. Deploying the jib results in a gross capacity deduction of 250 lb.
   d. Exchanging the hook and ball for a load block results in an additional 140 lb deduction from gross capacity.

Summary

Crane operators must understand how to read and interpret load charts for every crane they operate. The range diagram section of load charts illustrates the allowable operating radius, boom-tip heights, and boom angles possible with the crane in any allowed configuration. The load charts are the first thing operators should reference when planning a lift. The notes and specifications in the charts provide a wealth of information related to crane capacity and operation. The capacity tables of load charts provide the gross capacity for a given crane configuration.

Manufacturers establish the gross capacities listed on the charts under ideal lifting conditions and well within structural and tipping limits. The purpose of the charts is to provide the operator with reliable and authoritative guidance to operate the crane safely. Although operators will become thoroughly familiar with the charts of a crane after a period of use, they should never rely on memory with so much at stake. A crane operator must study and use the load charts for every new situation to ensure that every lift is within the capabilities of the equipment.

1. The point where you can assume an object's weight to be located is the _____.

   a. moment
   b. center of gravity
   c. tipping axis
   d. gross weight

2. If a crane's load chart becomes lost or illegible, _____.

   a. only the manufacturer can replace it
   b. the company engineer can recreate it
   c. an OSHA-qualified person may reproduce it
   d. the company must sell the associated crane for scrap

3. Which of the following items is no longer required to be included in crane load charts?

   a. Range chart showing allowed boom angles and load radius
   b. Work area chart
   c. Load limitations based on strength of materials, hydraulics, stability, etc.
   d. Cautions, warnings, and notes relating to load limitations

4. The quantity obtained from crane gross capacity minus weight deductions is _____.

   a. crane leverage
   b. crane net capacity
   c. load leverage
   d. net load

5. The boom of industrial/all-purpose cranes can usually rotate _____.

   a. 360 degrees
   b. 180 degrees
   c. 90 degrees
   d. 12 degrees (6 degrees to each side of center)

6. When operating a crane by the load charts (rather than an LMI), always _____.

   a. round the results of calculations down
   b. interpolate the chart data as needed
   c. enter the chart at the next highest load radius or next lowest boom angle
   d. operate the crane at less than 75 percent of the net capacity

7. An operational quadrant is defined _____.

   a. by the boom angle above the horizontal
   b. by the tipping axis for a given configuration and the direction of the crane boom
   c. by the front, side, and rear faces of the crane when on rubber or crawlers
   d. only when outriggers are in use

8. Load chart capacities for on-outrigger lifting do not exceed _____.

   a. 65 percent of the tipping loads
   b. 75 percent of the tipping loads
   c. 85 percent of the tipping loads
   d. 95 percent of the tipping loads

9. The offset of a jib and boom refers to _____.

   a. the angle formed by an erected jib in relation to the center line of the main boom
   b. how far off-center the jib is mounted on the end of the main boom
   c. the length of a boom extension between the main boom and the jib-boom itself
   d. the distance between the jib-boom tip and the end of the main boom

10. What is the industry standard for the ratio of a crane's hoist-line minimum breaking strength to its maximum single-line pull?

    a. 1:5
    b. 1:1
    c. 2:1
    d. 5:1

11. What information regarding crane configuration would you *not* find on a range diagram?

    a. Permitted crane boom configurations and angles
    b. Crane lift capacities for various boom angles and configurations
    c. Load radiuses for allowed boom configurations
    d. Telescopic boom-tip heights and boom angles at partial extensions for a given load radius

12. Referring to the Terex RT555-1 load charts in *Appendix A*, find the net capacity of the crane configured as follows:

- On outriggers, fully extended
- Lift over side
- 50-foot main boom length
- 22-foot load radius
- 4-sheave block on main boom (690 lb)
- Counterweights installed (included)
- Hook and ball (239 lb)
- Rigging and accessories (65 lb)
- Lifting from the main boom

a. 42,626 lb
b. 41,876 lb
c. 40,856 lb
d. 40,106 lb

13. Referring to the load charts for the Manitowoc 4100W-2 in *Appendix B*, find the *gross capacity* for the following configuration:

- All required counterweights installed, totaling 206,400 lb
- #22C open-throat top boom
- 180-foot main boom
- Boom angle is 60 degrees
- Crawlers fully extended

a. 37,900 lb
b. 40,900 lb
c. 45,300 lb
d. 52,100 lb

14. Referring to the load charts for the Manitowoc 4100W-2 in *Appendix B*, find the *net capacity* for the following configuration:

- All required counterweights installed, totaling 206,400 lb
- Crawlers fully extended
- #22C open-throat top boom; 190-ft main boom
- Lifting from the main boom hoist
- 130-ft load radius
- 2-sheave boom point attached
- 15-ton hook block (on whip line), 20 ft below boom point
- 100-ton block reeved with 4 parts of line
- 345 lb of rigging

a. 17,219 lb
b. 18,209 lb
c. 18,278 lb
d. 19,029 lb

15. Referring to the load charts for the Grove TM-1500 truck-mounted telescopic-boom crane in *Appendix C*, find the net capacity for the following configuration:

- Outriggers properly set, fully extended, and level
- Over-the-side lift
- 33-foot fixed extension erected at 15-degree offset
- Boom length of 173 feet
- Load radius of 68 feet
- 30-ton, 1-sheave block
- Auxiliary boom head attached
- 10-ton headache ball reeved over auxiliary boom head
- 167 lb of rigging and other hoisting accessories

a. 10,460 lb
b. 10,660 lb
c. 10,840 lb
d. 10,940 lb

Trade Terms Introduced in This Module

Boom angle: In load charts, the angle between the boom relative to the horizontal plane. On telescopic-boom cranes, boom angle instruments indicate the angle between the bottom of the base section and the horizontal with the boom loaded. On lattice-boom cranes, it is the angle between the center line of the boom and a horizontal line.

Boom length: Measured from the boom foot pins to the center of the boom-point sheaves. It does not include the length of an attached jib.

Boom-point elevation: Distance from the ground to the center of the boom-point sheaves.

Configuration: A term that addresses all aspects of a crane's condition for a given lift that can include boom assembly and position, hoisting line and accessories, base arrangement, counterweights, and other factors addressed in a load chart.

Effective weight: With reference to jib/boom extensions, the weight calculated by the manufacturer which, when applied at the boom tip, has the same effect on the crane capacity as the jib and/or boom extension itself.

Gross capacity: A crane capacity value as displayed in the crane's load chart for a specified configuration.

Gross load: The total weight of all the items lifted by the crane. The gross capacity of the crane minus any deductions for the crane configuration results in the gross load that can be lifted. Gross load is directly related to net capacity.

Interpolate: To estimate or calculate an unknown intermediate value that falls between two known values.

Offset: The angle formed by an erected boom, measured from an imaginary line extending through the center line of the main boom. The crane's load chart provides capacities for allowable jib or boom-extension offset angles.

Line pull: The maximum pulling force a crane hoist drum can apply to a hoist cable, accounted for by the torque that the motor can apply to the drum and the working radius of the rope as it leaves the drum.

Load operating radius: The horizontal distance from a vertical line through the axis of crane rotation to the center of the vertical hoist line or tackle, with a load applied. Also called load radius, lifting radius, or operating radius.

Net capacity: The crane's hoisting force available after taking into account the crane's and/or load's configurations relating to a lift. Equal to load chart gross capacity minus weight deductions.

Ringer crane: A large, high-capacity crane erected on a circular track (the ring, usually installed directly on the ground) that supports the entire crane base.

Tipping axis: In crane operations, an imaginary horizontal line around which a crane's center of gravity would rotate as it tips.

TEREX RT555-1 LOAD CHART

Lifting Capacities – Pounds (33' – 110' boom)

> **⚠ CAUTION:** Do not use this specification sheet as a load rating chart. The format of data is not consistent with the machine chart and may be subject to change.

MODEL RT555-1

COUNTERWEIGHT:	STABILITY PCT.
W/AUX. WINCH 13,100 LBS.	ON OUTRIGGERS 85%
W/O AUX. WINCH 14,200 LBS.	ON TIRES 75%
BOOM LENGTH 33-110 FT.	PCSA CLASS 10-210
OUTRIGGER SPREAD 22 FT.	

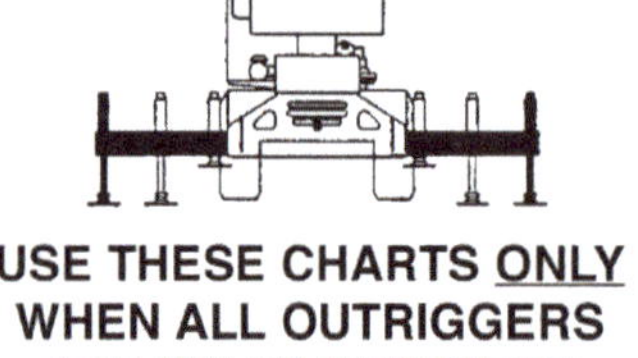

USE THESE CHARTS ONLY WHEN ALL OUTRIGGERS ARE FULLY EXTENDED

ON OUTRIGGERS - FULLY EXTENDED

LOAD RADIUS (FT)	BOOM LENGTH 35 FT			BOOM LENGTH 50 FT			BOOM LENGTH 65 FT			LOAD RADIUS (FT)
	BOOM ANGLE (DEG)	LOADED OVER FRONT (LB)	360° (LB)	BOOM ANGLE (DEG)	LOADED OVER FRONT (LB)	360° (LB)	BOOM ANGLE (DEG)	LOADED OVER FRONT (LB)	360° (LB)	
10	66.7	110,000*	110,000*	73.9	60,100*	60,100*				10
12	63.1	96,700*	93,700*	71.5	60,100*	60,100*				12
15	57.5	75,200*	73,100*	67.9	60,100*	60,100*	73.2	58,800*	58,800*	15
20	47.1	53,600*	52,300*	61.5	54,900*	53,600*	68.5	52,200*	52,200*	20
25	34.5	40,700*	39,700*	54.8	42,000*	41,100*	63.7	42,700*	41,700*	25
30	14.8	31,900*	31,200*	47.4	33,400*	32,700*	58.6	34,100*	33,400*	30
35	**			39.0	27,300*	26,700*	53.3	28,000*	27,400*	35
40				28.8	22,000	21,000	47.6	22,700	21,700	40
45				12.4	17,400	16,500	41.3	18,300	17,400	45
50				**			34.1	14,900	14,200	50
55							25.2	12,300	11,700	55
60							10.9	10,100	9,600	60
65							**			65
70										70
75										75
80										80
85										85
90										90
95										95
100										100
105										105
110										110

ON OUTRIGGERS - FULLY EXTENDED

LOAD RADIUS (FT)	BOOM LENGTH 80 FT			BOOM LENGTH 95 FT			BOOM LENGTH 110 FT			LOAD RADIUS (FT)
	LOADED BOOM ANGLE (DEG)	OVER REAR (LB)	360° (LB)	LOADED BOOM ANGLE (DEG)	OVER REAR (LB)	360° (LB)	LOADED BOOM ANGLE (DEG)	OVER REAR (LB)	360° (LB)	
10										10
12										12
15										15
20	72.7	38,700*	38,700*							20
25	68.9	33,600*	33,600*	72.3	29,300*	29,300*				25
30	65.0	29,600*	29,600*	69.1	25,900*	25,900*	72.1	22,900*	22,900*	30
35	61.0	26,500*	26,500*	65.9	23,000*	23,000*	69.3	20,500*	20,500*	35
40	56.8	23,000	22,000	62.5	20,800*	20,800*	66.5	18,400*	18,400*	40
45	52.4	18,600	17,700	59.1	18,800	17,900	63.6	16,500*	16,500*	45
50	47.7	15,300	14,600	55.5	15,500	14,800	60.7	14,900*	14,900	50
55	42.7	12,700	12,100	51.7	12,900	12,300	57.7	13,000	12,400	55
60	37.1	10,700	10,100	47.8	10,900	10,400	54.5	11,000	10,500	60
65	30.6	9,000	8,500	43.6	9,200	8,800	51.3	9,400	8,900	65
70	22.6	7,500	7,100	39.0	7,900	7,400	47.8	8,000	7,600	70
75	9.8	6,300	5,900	33.9	6,700	6,300	44.2	6,800	6,500	75
80	**			28.1	5,700	5,300	40.4	5,900	5,500	80
85				20.8	4,800	4,400	36.1	5,000	4,700	85
90				9.0	3,900	3,600	31.5	4,200	3,900	90
95				**			26.5	3,500	3,200	95
100							19.3	2,900	2,400	100
105							8.4	2,300	2,100	105
110							**			110

** MAXIMUM CAPACITY AT 0 DEGREE BOOM ANGLE

BOOM LENGTH 35 FT			BOOM LENGTH 50 FT			BOOM LENGTH 65 FT			BOOM LENGTH 80 FT			BOOM LENGTH 95 FT			BOOM LENGTH 110 FT		
LOAD RADIUS (FT)	OVER FRONT (LB)	360° (LB)	LOAD RADIUS (FT)	OVER FRONT (LB)	360° (LB)	LOAD RADIUS (FT)	OVER FRONT (LB)	360° (LB)	LOAD RADIUS (FT)	OVER FRONT (LB)	360° (LB)	LOAD RADIUS (FT)	OVER FRONT (LB)	360° (LB)	LOAD RADIUS (FT)	OVER FRONT (LB)	360° (LB)
31.2	20,900*	20,800*	46.2	12,600*	12,700*	61.2	8,200*	8,200*	76.2	5,400*	5,400*	91.2	3,500*	3,300	106.17	2,100	1,800

GENERAL NOTES

GENERAL

1. Rated loads as shown on Lift Charts pertain to this machine as originally manufactured and equipped. Modifications to the machine or use of optional equipment other than that specified can result in a reduction of capacity.

2. Construction equipment can be hazardous if improperly operated or maintained. Operation and maintenance of this machine shall be in compliance with the information in the Operator's, Parts and Safety Manuals supplied with this machine. If these manuals are missing, order replacements from the manufacturer through your distributor.

3. These warnings do not constitute all of the operating conditions for the crane. The operator and job site supervision must read the OPERATORS MANUAL, CIMA SAFETY MANUAL, APPLICABLE OSHA REGULATIONS, AND SOCIETY OF MECHANICAL ENGINEERS (ASME) SAFETY STANDARDS FOR CRANES.

4. This crane and its load ratings are in accordance with POWER CRANE & SHOVEL ASSOCIATION, STANDARD NO. 4, SAE CRANE LOAD STABILITY TEST CODE J765A, SAE METHOD OF TEST FOR CRANE STRUCTURE J1063 AND APPLICABLE SAFETY CODE FOR CRANES, DERRICKS AND HOISTS, ASME/ANSI B30.5.

DEFINITIONS

1. LOAD RADIUS – The horizontal distance from the axis of rotation before loading to the center of the vertical hoist line or tackle with a load applied.

2. LOADED BOOM ANGLE – It is the angle between the boom base section and the horizontal, after lifting the rated load at the rated radius. The boom angle before loading should be greater to account for deflections. The loaded boom angle combined with boom length give only an approximation of the operating radius.

3. WORKING AREA – Areas measured in a circular arc about the centerline of rotation as shown in the diagram.

4. FREELY SUSPENDED LOAD – Load hanging free with no direct external force applied except by the hoist rope.

5. SIDE LOAD – Horizontal force applied to the lifted load either on the ground or in the air.

6. NO LOAD STABILITY LIMIT – The stability limit radius shown on the range diagrams is the radius beyond which it is not permitted to position the boom, when the boom angle is less than the minimum shown on the applicable load chart, because the machine can overturn without any load.

7. BOOM SIDE OF CRANE – The side of the crane over which the boom is positioned when in an OVER SIDE working position.

SET–UP

1. Crane load ratings are based on the crane being leveled and standing on a firm, uniform supporting surface.

2. Crane load ratings on outriggers are based on all outrigger beams being fully extended or in the case of partial extension ratings mechanically pinned in the appropriate position, and the tires free of the supporting surface.

3. Crane load ratings on tires depend on appropriate inflation pressure and the tire conditions. Caution must be exercised when increasing air pressures in tires. Consult Operator's Manual for precautions.

4. Use of jibs, lattice–type boom extensions, or fourth section pullouts extended is not permitted for pick and carry operations.

5. Consult appropriate section of the Operator's and Service Manual for more exact description of hoist line reeving.

6. The use of more parts of line than required by the load may result in having insufficient rope to allow the hook block to reach the ground.

7. Properly maintained wire rope is essential for safe crane operation. Consult Operator's Manual for proper maintenance and inspection requirements.

8. When spin-resistant wire rope is used, the allowable rope loading shall be the breaking strength divided by five (5), unless otherwise specified by the wire rope manufacturer.

9. Do not elevate the boom above 60° unless the boom is positioned in-line with the crane's chassis or the outriggers are extended. Failure to observe this warning may result in loss of stability.

OPERATION

1. CRANE LOAD RATINGS MUST NOT BE EXCEEDED. DO NOT ATTEMPT TO TIP THE CRANE TO DETERMINE ALLOWABLE LOADS.

2. When either radius or boom length, or both, are between listed values, the smaller of the two listed load ratings shall be used.

3. Do not operate at longer radii than those listed on the applicable load rating chart (cross hatched areas shown on range diagrams).

4. The boom angles shown on the Capacity Chart give an approximation of the operating radius for a specified boom length. The boom angle, before loading, should be greater to account for boom deflection. It may be necessary to retract the boom if maximum boom angle is insufficient to maintain rated radius.

5. Power telescoping boom sections must be extended equally.

6. Rated loads include the weight of hook block, slings, and auxiliary lifting devices. Their weights shall be subtracted from the listed rated load to obtain the net load that can be lifted.

   When lifting over the jib the weight of any hook block, slings, and auxiliary lifting devices at the boom head must be added to the load.

   When jibs are erected but unused add two (2) times the weight of any hook block, slings, and auxiliary lifting devices at the jib head to the load.

7. Rated loads do not exceed 85% on outriggers or 75% on tires, of the tipping load as determined by SAE Crane Stability Test Code J765a. Structural strength ratings in chart are indicated with an asterisk (*).

8. Rated loads are based on freely suspended loads. No attempt shall be made to drag a load horizontally on the ground in any direction.

9. The user shall operate at reduced ratings to allow for adverse job conditions, such as: Soft or uneven ground, out of level conditions, high winds, side loads, pendulum action, jerking or sudden stopping of loads, hazardous conditions, experience of personnel, two machine lifts, traveling with loads, electric wires, etc., (side pull on boom or jib is hazardous). Derating of the cranes lifting capacity is required when wind speed exceeds 20 MPH. the center of the lifted load must never be allowed to move more than 3* feet off the center line of the base boom section due to the effects of wind, inertia, or any combination of the two.

   *"Use 2 feet off the center line of the base boom for a two section boom, 3 feet for a three section boom, or 4 feet for a four section boom."

10. The maximum load which can be telescoped is not definable, because of variations in loadings and crane maintenance, but it is permissible to attempt retraction and extension if load ratings are not exceeded.

11. Load ratings are dependent upon the crane being maintained according to manufacturer's specifications.

12. It is recommended that load handling devices, including hooks, and hook blocks, be kept away from boom head at all times.

13. FOR TRUCK CRANES ONLY: 360° capacities apply only to machines equipped with a front outrigger jack and all five (5) outrigger jacks properly set. If the front (5th) outrigger jack is not properly set, the work area is restricted to the over side and over rear areas as shown on the Crane Working Positions diagram. Use the 360° load ratings in the overside work areas.

14. Do not lift with outrigger beams positioned between the fully extended and intermediate (pinned) positions.

15. Truck Cranes <u>not</u> equipped with equalizing (bogie) beams between the rear axles may not be used for lifting "on tires". Truck Cranes equipped with equalizing beams and rear air suspension should "dump" the air before lifting "on tires".

CLAMSHELL, MAGNET, AND CONCRETE BUCKET SERVICE

1. Maximum boom length for clamshell and magnet service is 50 feet.

2. Weight of clamshell or magnet, plus contents are not to exceed 6,000 pounds or 90% of rated lifting capacities, whichever is less. For concrete bucket operation, weight of bucket and load must not exceed 90% of rated lifting capacity.

WE RESERVE THE RIGHT TO AMEND THESE SPECIFICATIONS AT ANY TIME WITHOUT NOTICE. THE ONLY WARRANTY APPLICABLE IS OUR STANDARD WRITTEN WARRANTY APPLICABLE TO THE PARTICULAR PRODUCT AND SALE. WE MAKE NO OTHER WARRANTY. EXPRESSED OR IMPLIED.

TEREX Cranes
106 12th Street S.E.
Waverly, IA 50677-9466 USA
TEL: +1 (319) 352-3020
FAX: +1 (319) 352-5727
EMAIL: inquire@terexwaverly.com
WEB: terex.com

MANITOWOC 4100W-2 LOAD CHART

WEIGHTS	4100W-S1 4100W-S2

DESCRIPTION	APPROX. WEIGHT (IN LBS.)

LIFTCRANE- w/70' No. 22C boom, machine counterweight, universal gantry w/telescopic backhitch, 26'-6" crawlers w/48" treads, full width tandem drums, ind. swing, ind. boom hoist, Cummins NTA-855-C360 engine, 27-5/8" dia. lagging for rear drum, boom hoist rope, equalizer, telescopic air cushioned boom stop, 200 ton load block, 15 ton hook and weight ball, single sheave upper boom point, and upper wire rope guide............................ S1 363,795*
S2 446,295*

UPPERWORKS- w/Cummins NTA-855-C360 engine, ind. boom hoist, ind. swing, and 27-5/8" dia. lagging for rear drum; <u>LESS</u> boom, gantry and backhitch, equalizer, load block, weight ball, counter weights, telescopic air cushioned boom stop, upper wire rope guide, and catwalk.................... 80,500*

UPPERWORKS as above - w/gantry and backhitch, equalizer, boom hoist rope, and carbody; <u>LESS</u> crawlers.. 138,685*

CARBODY-w/roller path, ring gear, and king pin; <u>LESS</u> crawlers........................... 49,700

CRAWLERS, 26'-6" w/48" treads.. 37,965 each

COUNTERWEIGHT

Inner (Self-Removing)..	41,900
Middle (Self-Removing)...	41,500
Outer (Self-Removing)...	39,000
Side (2 Req'd)...	12,000 each
Carbody (2 Req'd)...	30,000 each

BOOM No. 22C

Boom Butt - 30'...	6,150
Boom Top - 40' (w/Lower boom point assembly)..	8,445
Upper Boom Point (Removable - single sheave)..	1,260
double sheave...	1,505
Jib Adapter (Removable)...	545
Boom Insert - 10' (w/rope guide roller assembly)......................................	1,350
Boom Insert - 20' (w/rope guide roller assembly)......................................	2,435
Boom Insert - 40' (w/rope guide roller assembly)......................................	4,460
Boom Insert - 40' (w/jib backstay, and rope guide roller assembly)...................	4,560
Basic Pendant - 40' 9-3/4" (4 Req'd)...	255 each
Pendant - 10' (4 per insert)...	115 each
Pendant - 20' (4 per insert)...	155 each
Pendant - 40' (4 per insert)...	215 each
Pendant Spreader Bar..	320

*Weights do not include hoist line, whip line, or fuel. For CAT. D-343TA add 1,170 lbs., for CAT. 3406 PCTA add 100 lbs. and for CM 12 V-71N engines add 600 lbs.

DESCRIPTION	APPROX. WEIGHT (IN LBS.)

JIB NO. 123

Jib Top - 15' (w/Jib point)	695
Jib Butt - 15'	690
Jib Insert - 10'	340
Basic Pendant - 33' 3-3/4" (2 Req'd)	115 each
Pendant - 10' (2 Per insert)	65 each
Jib Backstay Pendant	155 each
Jib Strut - 12' - 6"	365

COMPONENTS

Hook Rollers (6) - w/shafts	1,020
Light Plant - 6.5KW - w/mounting platform	1,390
Catwalk - left and right side w/rails	1,320
Lagging - 27-5/8" dia. plain	1,410
Boom Hoist Rope - 12 Part - 760' of 7/8" - 6 x 26	1,080
Wire Rope Guide Assembly - Lower	325
Wire Rope Guide Assembly - Upper	510
Rope Guide Roller Assembly	55 each
2-Part Gantry w/Telescopic Backhitch	7,805
Equalizer	2,000
Hoist Line - 1-1/8" - 6 x 31	2.34 lbs./ft.
Whip Line - 1-1/8" - 6 x 31	2.34 lbs./ft.
15-Ton Hook and Weight Ball	865
100-Ton Hook Block Assembly	2,065
200-Ton Hook Block Assembly	4,900
230-Ton Hook Block Assembly	5,375
Boom Stop - Telescopic Air Cushioned	675
Dragline Fairlead - Revolving	1,910
Dragline Fairlead - Hinged	9,330

NOTE: The above weights may fluctuate up or down 5% due to manufacturing tolerances.

MANITOWOC 4100 W-2

LIFTCRANE CAPACITIES

BOOM NO. 22C WITH OPEN THROAT TOP
146,400 LB. CRANE COUNTERWEIGHT
60,000 LB. CARBODY COUNTERWEIGHT
26'6" CRAWLERS EXTENDED

WARNING: This chart will apply only when two 12,000 lb. side ctwts. and two 30,000 lb. carbody ctwts. bear MEC registered Serial Numbers.

LIFTING CAPACITIES: Capacities for various boom lengths and operating radii may be based on percent of tipping, strength of structural components, operating speeds and other factors.

Capacities are for freely suspended loads and do not exceed 75% of a static tipping load. Capacities based on structural competence are shown by shaded areas.

Capacities are shown in pounds. Deduct 1200 pounds from capacities listed when single sheave upper boom point is attached and 1500 pounds when two sheave upper boom point is attached. To comply with B30.5 requirements, upper boompoint cannot be used on the 260 ft. boom. Weight of jib, (see chart A), all load blocks, hooks, weight ball, slings, hoist lines beneath boom and jib point sheaves, etc., is considered part of the main boom load. Boom is not to be lowered beyond radii where combined weights are greater than rated capacity. Where no capacity is shown, operation is not intended or approved.

OPERATING CONDITIONS: Machine to operate in a level position on a firm surface with crawlers fully extended and gantry in working position and be rigged in accordance with and under conditions referred to in rigging drawing No. 190693 and load line specification chard No. 6592-A.

Crane operator judgement must be used to allow for dynamic load effects of swinging, hoisting or lowering, travel, as well as adverse operating conditions & physical machine depreciation.

OPERATOR RADIUS: Operating is the horizontal distance from the axis of rotation to the center of vertical hoist line or load block with the load freely suspended. Add 14" to boom point radius for radius of sheave when using single part hoist line.

Boom angle is the angle between horizontal and centerline of boom butt and inserts and is an indication of operating radius. In all cases, operating radius shall govern capacity.

BOOM POINT ELEVATION: Boom point elevation, in feet, is the vertical distance from ground level to centerline of boom point shaft.

MACHINE EQUIPMENT: Machine equipped with 26'6" extendible crawlers, 48" treads, 17' retractable gantry, 12 part boom hoist reeving, four 1 3/8" boom pendants, 1st ctwt. 41,900 lbs., 2nd ctwt. 41,500 lbs., 3rd ctwt. 39,000 lbs., two 12,000 lbs. side ctwt's. and two 30,000 lbs. carbody ctwt's.

LOAD AND WHIP LINE SPECIFICATIONS	
LOAD LINE:	1-1/8" - 6 x 31 Warrington-Seale, Extra Improved Plow Steel, Regular Lay, IWRC. Minimum Breaking Strength 65 Ton. (Approx.Weight Per Ft. in Lbs. 2.34)
WHIP LINE:	1-1/8" - Warrington-Seale, Improved Plow Steel, Regular Lay, IWRC. Minimum Breaking Strength 56.5 Ton. Maximum Load - 28,300 Lbs. Per Line. (Approx. Weight Per Ft. in Lbs. 2.34)

HOIST REEVING FOR MAIN LOAD BLOCK

No.Parts Of Line	1	2	3	4	5	6
Max Load - Lbs.	32,500	65,000	97,500	130,000	162,500	195,000
No. Parts of Line	7	8	9	10	11	12
Max. Load - Lbs.	227,500	260,000	292,500	325,000	357,500	400,000
No. Parts of Line	13					
Max. Load - Lbs.	430,000					

MAXIMUM BOOM AND JIB LENGTH LIFTED UNASSISTED				DEDUCT FROM CAPACITIES WHEN JIB IS ATTACHED	
OVER FRONT OF BLOCKED CRAWLERS		OVER SIDE OF EXTENDED CRAWLERS			
BOOM LENGTH	JIB NO. 123	BOOM LENGTH	JIB NO. 123	JIB LENGTH	JIB NO. 123
260'	--	260'	--	30'	3,000 lbs.
250'	--	250'	--	40'	3,600 lbs.
240'	40'	240'	40'	50'	4,200 lbs.
230'	60'	230'	60'	60'	4,900 lbs.

Load block, hook and weight ball on ground to start.

FOR JIB CAPACITIES, CONSULT JIB CHART.

BOOM LGTH FEET	OPER. RAD. FEET	BOOM ANG. DEG.	BOOM POINT ELEV.	CAPACITY: CRAWLERS EXTENDED
70	16.5	79.7	75.9	460,000
	17	79.3	75.8	400,000
	18	78.5	75.6	380,100
	19	77.6	75.4	363,000
	20	76.8	75.1	347,300
	22	75.1	74.6	319,600
	24	73.4	74.1	293,400
	26	71.7	73.5	266,100
	28	69.9	72.8	237,500
	30	68.2	72.0	214,300
	32	66.4	71.2	195,100
	34	64.6	70.2	178,900
	36	62.8	69.3	165,200
	38	60.9	68.2	153,300
	40	59.1	67.0	143,000
	45	54.1	63.7	122,100
	50	48.9	59.8	106,300
	55	43.2	54.9	93,900
	60	36.9	49.0	84,000
	65	29.4	41.3	75,800
	70	19.5	30.3	63,900

BOOM LGTH FEET	OPER. RAD. FEET	BOOM ANG. DEG.	BOOM POINT ELEV.	CAPACITY: CRAWLERS EXTENDED
80	17	80.6	85.9	392,800
	18	79.9	85.8	378,900
	19	79.2	85.6	361,800
	20	78.5	85.4	346,100
	22	77.0	84.9	318,400
	24	75.5	84.5	292,500
	26	74.0	83.9	265,600
	28	72.5	83.3	237,000
	30	71.0	82.7	213,800
	32	69.5	81.9	194,600
	34	68.0	81.2	178,500
	36	66.4	80.3	164,700
	38	64.8	79.4	152,800
	40	63.3	78.4	142,400
	45	59.2	75.7	121,500
	50	54.9	72.5	105,700
	55	50.4	68.6	93,300
	60	45.6	64.1	83,400
	65	40.3	58.8	75,200
	70	34.4	52.2	68,300
	75	27.4	43.9	52,500
	80	18.2	32.0	53,900

BOOM LGTH FEET	OPER. RAD. FEET	BOOM ANG. DEG.	BOOM POINT ELEV.	CAPACITY: CRAWLERS EXTENDED
90	18	81.1	95.9	355,400
	19	80.4	95.7	346,900
	20	79.8	95.6	336,900
	22	78.5	95.2	317,400
	24	77.2	94.7	291,700
	26	75.9	94.3	264,800
	28	74.5	93.7	236,600
	30	73.2	93.2	213,400
	32	71.9	92.5	194,200
	34	70.5	91.9	178,000
	36	69.2	91.1	164,200
	38	67.8	90.3	152,300
	40	66.4	89.5	142,000
	45	62.9	87.1	121,100
	50	59.3	84.4	105,200
	55	55.5	81.2	92,800
	60	51.5	77.5	82,900
	65	47.3	73.2	74,700
	70	42.8	68.2	67,800
	75	37.9	62.3	62,000
	80	32.4	55.2	57,000
	85	25.8	46.2	52,600
	90	17.1	33.5	45,900

BOOM LGTH FEET	OPER. RAD. FEET	BOOM ANG. DEG.	BOOM POINT ELEV.	CAPACITY: CRAWLERS EXTENDED
100	19	81.4	105.9	332,900
	20	80.8	105.7	327,100
	22	79.6	105.4	316,200
	24	78.5	105.0	290,800
	26	77.3	104.5	263,900
	28	76.1	104.1	236,200
	30	74.9	103.6	212,900
	32	73.7	103.0	193,700
	34	72.5	102.4	177,500
	36	71.3	101.7	163,700
	38	70.1	101.0	151,800
	40	68.9	100.3	141,400
	45	65.8	98.2	120,500
	50	62.6	95.8	104,700
	55	59.3	93.0	92,300
	60	55.9	89.8	82,300
	65	52.4	86.2	74,100
	70	48.7	82.1	67,200
	75	44.8	77.4	61,400
	80	40.5	72.0	56,400
	85	35.9	65.6	52,000
	90	30.7	58.0	48,200
	95	24.5	48.5	44,900
	100	16.3	35.0	39,300

CAUTION! CHECK AMOUNT OF COUNTERWEIGHT ON MACHINE BEFORE USE OF THIS CHART

NCCER – *Advanced Rigger*

BOOM LGTH FEET	OPER. RAD. FEET	BOOM ANG. DEG.	BOOM POINT ELEV.	CAPACITY: CRAWLERS EXTENDED
110	22	82.3	115.7	240,500
	24	81.3	115.4	232,100
	26	80.2	115.0	224,400
	28	79.1	114.6	217,300
	30	78.1	114.1	210,700
	32	77.0	113.6	193,300
	34	75.9	113.1	177,100
	36	74.8	112.5	163,300
	38	73.7	111.8	151,400
	40	72.6	111.2	141,000
	45	69.9	109.3	120,100
	50	67.0	107.2	104,200
	55	64.1	104.7	91,800
	60	61.2	101.9	81,800
	65	58.1	98.8	73,600
	70	54.9	95.3	66,800
	75	51.6	91.3	60,900
	80	48.1	86.8	55,900
	85	44.4	81.7	51,600
	90	40.4	75.9	47,800
	95	36.1	69.2	44,400
	100	31.1	61.1	41,400
	105	25.3	51.1	38,700
	110	17.6	37.2	33,900
120	22	83.0	125.9	228,700
	24	82.0	125.5	220,500
	26	81.0	125.2	213,000
	28	80.0	124.8	206,100
	30	79.1	124.4	199,700
	32	78.1	123.9	192,800
	34	77.1	123.4	176,600
	36	76.1	122.9	162,800
	38	75.1	122.3	150,900
	40	74.1	121.7	140,500
	45	71.6	120.0	119,500
	50	69.1	118.1	103,700
	55	66.5	115.9	91,300
	60	63.8	113.4	81,300
	65	61.1	110.6	73,100
	70	58.3	107.5	66,200
	75	55.4	104.0	60,400
	80	52.4	100.1	55,300
	85	49.2	95.8	51,000
	90	45.9	91.0	47,200
	95	42.4	85.5	43,800
	100	38.6	79.4	40,800
	105	34.4	72.2	38,100
	110	29.7	63.7	35,700
	115	24.1	53.2	33,500
130	24	82.6	135.7	212,700
	26	81.7	135.3	205,300
	28	80.8	135.0	198,600
	30	79.9	134.6	192,300
	32	79.0	134.2	186,500
	34	78.1	133.7	176,300
	36	77.2	133.2	162,500
	38	76.3	132.7	150,600
	40	75.4	132.1	140,200
	45	73.1	130.6	119,300
	50	70.8	128.8	103,400
	55	68.4	126.8	91,000
	60	66.0	124.6	81,000
	65	63.5	122.0	72,800
	70	61.0	119.2	65,900
	75	58.4	116.1	60,000
	80	55.7	112.7	55,000
	85	53.0	108.9	50,700
	90	50.1	104.8	46,800
	95	47.1	100.1	43,500
	100	43.9	95.0	40,500
	105	40.6	89.2	37,800
	110	36.9	82.6	35,300
	115	33.0	75.1	33,100
	120	28.5	66.1	31,100
	125	23.1	55.1	29,300
140	26	82.3	145.5	198,400
	28	81.5	145.1	191,800
	30	80.7	144.8	185,600
	32	79.8	144.4	179,900
	34	79.0	144.0	174,700
	36	78.1	143.5	162,000
	38	77.3	143.0	150,100
	40	76.5	142.5	139,700
	45	74.3	141.1	118,700
	50	72.2	139.4	102,800
	55	70.0	137.6	90,400
	60	67.8	135.5	80,400
	65	65.6	133.2	72,200
	70	63.3	130.7	65,300
	75	60.9	127.9	59,500
	80	58.5	124.8	54,400
	85	56.1	121.4	50,100
	90	53.5	117.7	46,200
	95	50.9	113.7	42,900
	100	48.1	109.2	39,900

BOOM LGTH FEET	OPER. RAD. FEET	BOOM ANG. DEG.	BOOM POINT ELEV.	CAPACITY: CRAWLERS EXTENDED
140	105	45.3	104.2	37,200
	110	42.2	98.8	34,700
	115	39.0	92.7	32,500
	120	35.5	85.7	30,500
	125	31.7	77.8	28,700
	130	27.4	68.5	27,000
	135	22.2	57.0	25,500
150	28	82.1	155.3	185,600
	30	81.3	154.9	179,600
	32	80.5	154.6	174,000
	34	79.7	154.2	168,800
	36	79.0	153.8	161,600
	38	78.2	153.3	149,600
	40	77.4	152.8	139,200
	45	75.4	151.5	118,200
	50	73.4	150.0	102,400
	55	71.4	148.3	89,900
	60	69.4	146.4	79,900
	65	67.3	144.3	71,700
	70	65.2	141.9	64,800
	75	63.0	139.4	59,000
	80	60.9	136.6	53,900
	85	58.6	133.5	49,600
	90	56.3	130.1	45,700
	95	54.0	126.5	42,400
	100	51.5	122.5	39,400
	105	49.0	118.2	36,700
	110	46.4	113.4	34,200
	115	43.6	108.2	32,000
	120	40.7	102.4	30,000
	125	37.6	96.0	28,200
	130	34.2	88.8	26,500
	135	30.5	80.5	25,000
	140	26.4	70.8	23,500
	145	21.4	58.8	22,200
160	28	82.6	165.4	179,800
	30	81.8	165.1	173,900
	32	81.1	164.7	168,400
	34	80.4	164.4	163,300
	36	79.7	164.0	158,600
	38	78.9	163.6	149,100
	40	78.2	163.1	138,700
	45	76.3	161.9	117,700
	50	74.5	160.5	101,800
	55	72.6	158.9	89,300
	60	70.7	157.1	79,300
	65	68.8	155.1	71,100
	70	66.8	153.0	64,200
	75	64.9	150.6	58,400
	80	62.9	148.0	53,300
	85	60.8	145.2	49,000
	90	58.7	142.2	45,100
	95	56.6	138.8	41,800
	100	54.4	135.2	38,700
	105	52.1	131.4	36,000
	110	49.7	127.1	33,600
	115	47.3	122.5	31,400
	120	44.8	117.5	29,400
	125	42.1	112.0	27,600
	130	39.3	105.9	25,900
	135	36.3	99.2	24,300
	140	33.1	91.7	22,900
	145	29.5	83.0	21,600
	150	25.5	72.9	20,400
	155	20.7	60.5	19,200
170	30	82.3	175.2	169,000
	32	81.6	174.9	163,600
	34	81.0	174.5	158,700
	36	80.3	174.2	154,000
	38	79.6	173.8	148,800
	40	78.9	173.3	138,400
	45	77.2	172.2	117,400
	50	75.4	170.9	101,500
	55	73.7	169.4	89,100
	60	71.9	167.7	79,000
	65	70.1	165.9	70,800
	70	68.3	163.9	63,900
	75	66.4	161.7	58,100
	80	64.6	159.3	53,000
	85	62.7	156.7	48,700
	90	60.8	153.9	44,900
	95	58.8	150.8	41,500
	100	56.8	147.5	38,500
	105	54.7	144.0	35,800
	110	52.6	140.2	33,300
	115	50.4	136.0	31,100
	120	48.1	131.5	29,100
	125	45.8	126.7	27,300
	130	43.3	121.4	25,600
	135	40.8	115.7	24,100
	140	38.1	109.3	22,600
	145	35.2	102.3	21,300
	150	32.0	94.5	20,100
	155	28.6	85.5	18,900
	160	24.7	75.1	17,800
	165	20.0	62.2	16,400

BOOM LGTH FEET	OPER. RAD. FEET	BOOM ANG. DEG.	BOOM POINT ELEV.	CAPACITY: CRAWLERS EXTENDED
180	32	82.1	185.0	158,700
	34	81.5	184.7	154,000
	36	80.8	184.3	149,500
	38	80.2	184.0	145,200
	40	79.5	183.6	137,900
	45	77.9	182.5	116,900
	50	76.3	181.2	101,000
	55	74.6	179.8	88,500
	60	72.9	178.3	78,500
	65	71.3	176.6	70,300
	70	69.6	174.7	63,400
	75	67.8	172.6	57,500
	80	66.1	170.4	52,500
	85	64.3	168.0	48,100
	90	62.5	165.4	44,300
	95	60.7	162.5	40,900
	100	58.9	159.5	37,900
	105	57.0	156.2	35,200
	110	55.0	152.7	32,700
	115	53.0	149.0	30,500
	120	51.0	144.9	28,500
	125	48.9	140.5	26,700
	130	46.7	135.8	25,000
	135	44.4	130.7	23,500
	140	42.0	125.2	22,000
	145	39.6	119.2	20,700
	150	36.9	112.6	19,500
	155	34.1	105.3	18,300
	160	31.1	97.2	17,200
	165	27.7	87.9	16,200
	170	23.9	77.1	15,300
	175	19.4	63.8	13,500
190	32	82.5	195.1	153,100
	34	81.9	194.8	148,800
	36	81.3	194.5	144,800
	38	80.7	194.1	140,800
	40	80.1	193.8	137,000
	45	78.5	192.7	116,400
	50	77.0	191.5	100,500
	55	75.4	190.2	88,000
	60	73.9	188.8	78,000
	65	72.3	187.1	69,700
	70	70.7	185.4	62,800
	75	69.1	183.4	57,000
	80	67.4	181.4	52,000
	85	65.8	179.1	47,600
	90	64.1	176.6	43,800
	95	62.4	174.0	40,400
	100	60.7	171.2	37,400
	105	58.9	168.2	34,700
	110	57.1	164.9	32,200
	115	55.3	161.4	30,000
	120	53.4	157.7	28,000
	125	51.5	153.7	26,200
	130	49.5	149.5	24,500
	135	47.5	144.9	22,900
	140	45.3	140.0	21,500
	145	43.1	134.6	20,200
	150	40.8	128.9	18,900
	155	38.4	122.6	17,800
	160	35.9	115.8	16,700
	165	33.2	108.3	15,700
	170	30.2	99.8	14,800
	175	27.0	90.3	13,900
	180	23.3	79.1	13,100
	185	18.9	65.4	11,000
200	34	82.3	204.9	143,100
	36	81.7	204.6	139,500
	38	81.2	204.3	136,000
	40	80.6	203.9	132,500
	45	79.1	203.0	115,900
	50	77.7	201.8	100,000
	55	76.2	200.6	87,500
	60	74.7	199.2	77,400
	65	73.2	197.7	69,200
	70	71.7	196.0	62,300
	75	70.2	194.2	56,400
	80	68.6	192.2	51,400
	85	67.1	190.1	47,000
	90	65.5	187.8	43,200
	95	63.9	185.3	39,800
	100	62.3	182.7	36,800
	105	60.6	179.8	34,100
	110	59.0	176.8	31,600
	115	57.3	173.6	29,400
	120	55.5	170.2	27,400
	125	53.0	166.5	25,600
	130	51.9	162.6	23,900
	135	50.1	158.4	22,300
	140	48.1	153.9	20,900
	145	46.2	149.1	19,600
	150	44.1	144.0	18,300
	155	42.0	138.4	17,200
	160	39.7	132.5	16,100
	165	37.4	126.0	15,100
	170	34.9	118.9	14,200

BOOM LGTH FEET	OPER. RAD. FEET	BOOM ANG. DEG.	BOOM POINT ELEV.	CAPACITY: CRAWLERS EXTENDED
200	175	32.3	111.1	13,300
	180	29.4	102.4	12,400
	185	26.2	92.5	11,700
	190	22.6	81.0	10,900
	195	18.4	67.0	8,500
210	36	82.1	214.7	134,200
	38	81.6	214.4	131,000
	40	81.0	214.1	127,900
	45	79.6	213.2	115,500
	50	78.3	212.1	99,600
	55	76.9	210.9	87,100
	60	75.4	209.6	77,100
	65	74.0	208.1	68,800
	70	72.6	206.6	61,900
	75	71.2	204.8	56,100
	80	69.7	203.0	51,000
	85	68.2	201.0	46,700
	90	66.7	198.8	42,800
	95	65.2	196.5	39,400
	100	63.7	194.0	36,400
	105	62.2	191.3	33,700
	110	60.6	188.5	31,300
	115	59.0	185.5	29,100
	120	57.4	182.3	27,100
	125	55.7	178.9	25,200
	130	54.1	175.2	23,500
	135	52.3	171.4	22,000
	140	50.6	167.3	20,600
	145	48.8	162.9	19,200
	150	46.9	158.2	18,000
	155	45.0	153.2	16,800
	160	43.0	147.9	15,800
	165	40.9	142.1	14,800
	170	38.7	135.9	13,800
	175	36.5	129.2	12,900
	180	34.0	121.9	12,100
	185	31.5	113.9	11,300
	190	28.7	104.9	10,600
	195	25.6	94.7	9,900
	200	22.1	82.9	8,800
	205	17.9	68.5	6,500
220	38	82.0	224.6	125,700
	40	81.4	224.2	122,800
	45	80.1	223.3	115,000
	50	78.8	222.3	99,100
	55	77.5	221.2	86,600
	60	76.1	220.0	76,600
	65	74.8	218.6	68,300
	70	73.4	217.1	61,400
	75	72.0	215.4	55,500
	80	70.7	213.7	50,500
	85	69.3	211.8	46,100
	90	67.9	209.7	42,300
	95	66.4	207.5	38,900
	100	65.0	205.2	35,800
	105	63.6	202.7	33,100
	110	62.1	200.0	30,700
	115	60.6	197.2	28,500
	120	59.1	194.2	26,500
	125	57.5	191.0	24,600
	130	55.9	187.6	23,000
	135	54.3	184.0	21,400
	140	52.7	180.2	20,000
	145	51.0	176.1	18,600
	150	49.3	171.8	17,400
	155	47.6	167.3	16,300
	160	45.7	162.4	15,200
	165	43.9	157.2	14,200
	170	41.9	151.7	13,200
	175	39.9	145.7	12,300
	180	37.8	139.3	11,500
	185	35.6	132.4	10,700
	190	33.2	124.8	10,000
	195	30.7	116.6	9,300
	200	28.0	107.3	8,600
	205	25.0	96.9	8,000
	210	21.5	84.8	6,500

CAPACITIES CONTINUED ON NEXT PAGE

These load charts are intended for instructional purposes only. They were derived from manufacturer sales information which may not be complete or machine specific. Not responsible for typographical errors.

GROVE TM1500 LOAD CHART

RATED LIFTING CAPACITIES IN POUNDS
46 FT. - 173 FT. BOOM ON OUTRIGGERS - OVER REAR

Radius in Feet	#0001									#0002
	Main Boom Length in Feet (Power Pinned Fly Retracted)									Power Pin Fly Ext. & 141 ft.
	46	58	70	82	94	106	118	130	141	173
10	300,000 (74.5)									
12	280,000 (72)	143,500 (76)	142,000 (79)							
15	235,000 (67.5)	143,500 (72.5)	141,500 (76.5)	130,000 (78.5)						
20	173,500 (60.5)	143,500 (67.5)	123,500 (72)	112,000 (75)	102,000 (77.5)	90,300 (79.5)				
25	136,500 (52)	131,500 (61.5)	110,500 (67.5)	98,650 (71)	89,250 (74)	78,550 (76.5)	73,700 (78.5)	69,300 (80)		
30	106,000 (43)	106,000 (55.5)	98,000 (63)	88,350 (67.5)	78,750 (71)	69,250 (73.5)	65,100 (75)	61,000 (77.5)	60,000 (79.5)	
35	84,700 (30.5)	84,700 (49)	84,700 (58)	80,150 (63.5)	69,000 (67.5)	60,750 (70.5)	57,150 (73)	54,000 (75.5)	52,150 (77.5)	
40		70,500 (41)	70,500 (52.5)	70,500 (59.5)	51,300 (64)	54,000 (67.5)	50,600 (70.5)	48,300 (73)	45,850 (75)	38,000 (79)
45		58,850 (32)	58,850 (47)	58,850 (55)	55,000 (60.5)	48,500 (64.5)	46,200 (68)	43,050 (71)	40,400 (73)	35,750 (77)
50		49,600 (17.5)	49,600 (40.5)	49,600 (50.5)	48,750 (57)	43,050 (61.5)	40,700 (65)	38,250 (68.5)	35,750 (71)	32,100 (75.5)
60			36,200 (22.5)	36,200 (38.5)	36,200 (48.5)	34,300 (55)	33,600 (58.5)	30,750 (63.5)	28,500 (66.5)	26,350 (72)
70				26,050 (25)	26,050 (39.5)	26,050 (47.5)	26,050 (53)	24,750 (58)	23,100 (61.5)	22,000 (68.5)
80					18,850 (27)	18,850 (39)	18,850 (46.5)	18,850 (52.5)	18,700 (56.5)	18,500 (64.5)
90						13,500 (28)	13,500 (38.5)	13,500 (46.5)	13,500 (51.5)	15,250 (60.5)
100							9,390 (29)	9,390 (39)	9,390 (45.5)	12,600 (56.5)
110							6,080 (12.5)	6,080 (30.5)	6,080 (39)	10,100 (52)
120								3,390 (17.5)	3,390 (31)	7,530 (47.5)
130									1,150 (19.5)	5,390 (42.5)
140										3,610 (36.5)
150										2,100 (30)
Minimum boom angle (deg.) for indicated length (no load)									10	19
Maximum boom length (ft) at 0 deg boom angle (no load)									140	167

NOTE: ( ) Boom angles are in degrees.
#LMI operating code. Refer to LMI manual for instructions.

A6-829-007136 & -007143

WEIGHT REDUCTIONS FOR LOAD HANDLING DEVICES

33 ft. Extension	
* Stowed -	892 lbs.
* Erected -	5,704 lbs.

33 ft. - 58 ft. Extension	
* Stowed -	1,246 lbs.
* Erected (Ret.) -	8,412 lbs.
* Erected (Ext.) -	11,406 lbs.

46 ft. - 173 ft. Boom with	
* 46 ft. Jib Erected -	9,613 lbs.
* 60 ft. Jib Erected -	14,571 lbs.
* 74 ft. Jib Erected -	20,443 lbs.
* 88 ft. Jib Erected -	27,199 lbs.
* Fixed Jib Accessories -	327 lbs.

* Reduction of main boom capacities

HOOKBLOCKS:	
30 Ton, 1 Sheave	1,022 lbs.
150 Ton, 8 Sheave	5,254 lbs.
Auxiliary Boom Head	261 lbs.
10 Ton Headache Ball	560 lbs.
15 Ton Headache Ball	803 lbs.

RATED LIFTING CAPACITIES IN POUNDS
46 FT. - 173 FT. BOOM ON OUTRIGGERS - 360°

Radius in Feet	#0001									#0002
	Main Boom Length in Feet (Power Pinned Fly Retracted)									Power Pin Fly Ext. & 141 ft.
	46	58	70	82	94	106	118	130	141	173
10	300,000 (74.5)									
12	280,000 (72)	143,500 (76)	142,000 (79)							
15	235,000 (67.5)	143,500 (72.5)	141,500 (76.5)	130,000 (78.5)						
20	173,500 (60.5)	143,500 (67.5)	123,500 (72)	112,000 (75)	102,000 (77.5)	90,300 (79.5)				
25	135,500 (52)	131,500 (61.5)	110,500 (67.5)	98,650 (71)	89,250 (74)	78,550 (76.5)	73,700 (78.5)	69,300 (80)		
30	106,000 (43)	106,000 (55.5)	98,000 (63)	88,350 (67.5)	78,750 (71)	69,250 (73.5)	65,100 (76)	61,000 (77.5)	60,000 (79.5)	
35	84,700 (30.5)	84,700 (49)	84,700 (58)	80,150 (63.5)	69,000 (67.5)	60,750 (70.5)	57,150 (73)	54,000 (75.5)	52,150 (77.5)	
40		70,500 (41)	70,500 (52.5)	70,500 (59.5)	61,300 (64)	54,000 (67.5)	50,600 (70.5)	48,300 (73)	45,850 (75)	38,000 (79)
45		57,250 (32)	57,250 (47)	57,250 (55)	55,000 (60.5)	48,500 (64.5)	45,200 (68)	43,050 (71)	40,400 (73)	35,750 (77)
50		46,550 (17.5)	46,550 (40.5)	46,550 (50.5)	46,550 (57)	43,050 (61.5)	40,700 (65)	38,250 (68.5)	35,750 (71)	32,100 (75.5)
60			32,000 (22.5)	32,000 (39.5)	32,000 (48.5)	32,000 (55)	32,600 (58.5)	30,750 (63.5)	28,500 (66.5)	26,350 (72)
70				22,550 (25)	22,550 (39.5)	22,550 (47.5)	22,550 (53)	22,550 (58)	22,550 (61.5)	22,000 (68.5)
80					15,900 (27)	15,900 (39)	15,900 (46.5)	15,900 (52.5)	15,900 (56.5)	18,500 (64.5)
90						11,000 (28)	11,000 (38.5)	11,000 (46.5)	11,000 (51.5)	15,250 (60.5)
100							7,280 (29)	7,280 (39)	7,260 (45.5)	11,750 (56.5)
110							4,260 (12.5)	4,260 (30.5)	4,260 (39)	8,520 (52)
120								1,830 (17.5)	1,830 (31)	5,850 (47.5)
130										3,610 (42.5)
140										1,710 (36.5)
Minimum boom angle (deg.) for indicated length (no load)									20	30
Maximum boom length (ft) at 0 dog boom angle (no load)									133	166

NOTE: ( ) Boom angles are in degrees.
#LMI operating code. Refer to LMI manual for instructions.

A6-829-007136 & -007143

33 FT. FIXED LENGTH EXTENSION ON OUTRIGGERS - 360°

Main Boom Angle (Deg.)	#0051		#0052		#0053	
	2° OFFSET		15° OFFSET		30° OFFSET	
	Rad. Ref. (ft.)	Cap. lbs.	Rad. Ref. (ft.)	Cap. lbs	Rad. Ref. (ft.)	Cap. lbs.
80	39.1	22,850	45.9	20,200	52.0	14,750
75	56.6	17,400	63.0	16,100	68.7	12,750
70	73.7	14,000	79.6	12,950	84.8	10,700
65	90.2	11,850	95.5	10,750	100.3	9,210
60	106.0	10,750	110.8	9,140	115.0	8,050
55	121.1	7,240	125.2	6,160	128.8	5,320
50	135.2	4,330	138.6	3,700	141.8	3,150

A6-829-008157

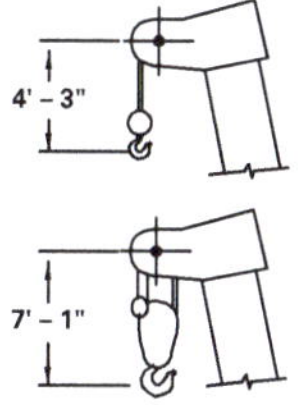

33 FT. - 58 FT. TELE EXTENSION ON OUTRIGGERS - 360°

Main Boom Angle (Deg.)	33 ft. EXTENSION						48 ft. EXTENSION						58 ft. EXTENSION					
	#0021		#0022		#0023		#0031		#0032		#0033		#0041		#0042		#0043	
	2° OFFSET		15° OFFSET		30° OFFSET		2° OFFSET		15° OFFSET		30° OFFSET		2° OFFSET		15° OFFSET		30° OFFSET	
	Rad. Ref. (ft.)	Cap. lbs.	Rad. Ref. (ft.)	Cap. lbs.	Rad. Ref. (ft.)	Cap. lbs.	Rad. Ref. (ft.)	Cap. lbs.	Rad. Ref. (ft.)	Cap. lbs.	Rad. Ref. (ft.)	Cap. lbs.	Rad. Ref. (ft.)	Cap. lbs.	Rad. Ref. (ft.)	Cap. lbs.	Rad. Ref. (ft.)	Cap. lbs.
80	39.1	22,300	45.9	19,650	52.0	14,200	43.9	14,800	55.8	12,700	65.0	9,040	45.1	9,600	59.1	8,300	69.6	7,000
75	56.6	16,850	63.0	15,550	68.7	12,200	62.5	13,700	73.4	10,900	81.8	8,320	65.0	9,150	77.9	7,950	87.8	6,440
70	73.7	13,450	78.6	12,400	84.8	10,150	80.7	11,600	90.3	8,840	97.9	7,110	84.4	8,550	96.2	7,200	105.5	5,760
65	90.2	11,300	95.5	10,200	100.3	8,870	98.2	9,230	106.7	7,330	113.3	6,120	103.2	7,630	113.8	6,000	122.3	4,970
60	106.0	10,200	110.8	8,600	115.0	7,450	115.1	7,550	122.1	8,210	127.9	5,340	121.2	6,260	130.5	5,100	138.2	4,380
55	121.1	6,400	125.2	5,370	128.8	4,540	131.1	5,200	136.7	3,860	141.5	2,950	138.2	4,650	146.2	3,265	153.1	2,380
50	135.2	3,520	138.6	2,810	141.8	2,360	146.1	2,580	150.2	1,790			154.3	2,235				

#LMI operating code. Refer to LMI manual for instructions.

A6-829-007158

Additional Resources

This module is intended as a thorough resource for task training. The following reference materials are recommended for further study.

ASME Standard B30.5, Mobile and Locomotive Cranes. Current edition. New York, NY: American Society of Mechanical Engineers.

Cranes: Design, Practice, and Maintenance, Ing J. Verschoof. 2002. Hoboken, NJ: John Wiley and Sons, Inc.

29 CFR 1926, Subpart CC, *Cranes and Derricks in Construction*. **www.ecfr.gov**

Figure Credits

Link-Belt Construction Equipment Company, Module opener, Figures 3, 4, 6A, 8, 11– 13, 16, 25, 27, 30, 31

Terex Cranes, used with permission of the owner, Figures 6B, 14, Appendix A

The Manitowoc Company, Inc., Figures 6C, 7, 9, 10, 18– 22, 28, 32, 34, 35, 37–40, 42, 43, 45, 46, Appendix B, Appendix C

TADANO America Corporation, Figure 15

Mammoet USA South Inc., Figures 23, 24, 26

© The Crosby Group LLC, Table 2

Broderson Manufacturing Corp., Exam art figure 1

Section Review Answer Key

Answer	Section Reference	Objective
Section One		
1. a	1.1.2	1a
2. d	1.2.0	1b
Section Two		
1. a	2.1.1	2a
2. b	2.2.0	2b
3. d	2.3.0	2c
4. a	2.4.0	2d
Section Three		
1. b	3.0.0	3a
2. d	3.2.1	3b
3. b	3.3.0	3c
4. b	3.4.1	3d

Lift Planning

OVERVIEW

Industrial and safety standards require some form of lift planning before a lift takes place. The level of detail required for a lift plan depends on the complexity of the lift, and the real and potential hazards involved. The more hazardous or sensitive the lift, the more detailed and formal the plan must be. A lift plan contains information relating to the crane(s) used, the load and its rigging, and all the site coordination issues. This module explains what lift planning involves and describes how to prepare a lift plan.

Module 21304

Objective

When you have completed this module, you will be able to do the following:

1. Describe lift planning, development, and implementation.
   a. Discuss the considerations that go into lift planning.
   b. Describe the development and contents of standard and critical lift plans.
   c. Describe how to implement a lift plan.

Performance Tasks

Under the supervision of your instructor, you should be able to do the following:

1. Perform a net-load capacity calculation for the crane specified in a scenario.
2. Fill in and complete a lift data sheet for a given lift scenario.

Trade Terms

Ground bearing pressure (GBP)
Lift data sheet (LDS)

Industry Recognized Credentials

If you are training through an NCCER-accredited sponsor, you may be eligible for credentials from NCCER's Registry. The ID number for this module is 21304. Note that this module may have been used in other NCCER curricula and may apply to other level completions. Contact NCCER's Registry at 888.622.3720 or go to **www.nccer.org** for more information.

Contents

1.0.0 LIFT PLANS

Objective

Describe lift planning, development, and implementation.

   a. Discuss the considerations that go into lift planning.
   b. Describe the development and contents of standard and critical lift plans.
   c. Describe how to implement a lift plan.

Performance Tasks

   1. Perform a net-load capacity calculation for the crane specified in a scenario.
   2. Fill in and complete a lift data sheet for a given lift scenario.

Trade Terms

Ground bearing pressure (GBP): The maximum force per unit area (pressure), measured in pounds per square foot, that can be exerted on the ground's surface without causing deformation of the surface. Can only be determined by testing.

Lift data sheet (LDS): The essential page or pages of a written lift plan that organize and summarize the important information required to develop a lift plan.

Crane operations are essential to most construction projects, but these operations carry great potential for accidents or errors. Mistakes can be costly, both monetarily and in terms of human lives or injuries. Lift planning, when properly done, identifies and addresses all the factors that can work against a safe and successful lift. Before a lift begins, relevant standards require that all workers involved minimize potential hazards as much as feasible.

1.1.0 Lift Planning Considerations

Every lift operation is unique in one way or another. However, crane lifts generally involve raising, lowering, or moving a load between two points. OSHA regulations and industry standards address lift planning and provide lift planning guidelines to enhance safety. This section identifies the key factors that lift planning should address.

1.1.1 Purposes and Standards

The primary purpose of lift planning standards and regulations is to protect lives and prevent injuries. The American Society of Mechanical Engineers (ASME) developed the most relevant and thorough lift-planning standard in *ASME P30.1, Planning for Load Handling Activities*. The sections that follow are summaries of the relevant content found in this standard. *Nonmandatory Appendix B—Industry References* of *ASME P30.1* provides a list of additional engineering standards that pertain to lift planning.

The Occupational Safety and Health Administration (OSHA) has also issued regulations that specifically address the safety aspects of lift planning. You can find this information in the relevant sections of 29 *CFR* Part 1926, Subpart CC, *Cranes and Derricks in Construction*.

Many employers and crane owners develop internal procedures that exceed OSHA requirements and *ASME P30.1* guidelines. It is essential to follow these procedures in addition to those outlined by OSHA and ASME.

> **NOTE**
>
> While planning for every lift should take into consideration the topics of the following sections, *ASME P30.1* states that documentation of the evaluation process is not required.

1.1.2 Lift Personnel and Responsibilities

ASME P30.1, Chapter 3—Personnel and Responsibilities lists the titles and responsibilities of personnel involved in crane operations. The employer must determine that these individuals are qualified and competent in order to ensure a safe and successful lifting operation. Not all roles listed here are needed for every operation, and a single individual may fulfill the responsibilities of one or more of the following positions:

- *Crane Assembly/Disassembly Director* – Directs the assembly, disassembly, or erection of the crane.
- *Crane Operator* – Directly controls the crane's functions.
- *Crane Owner* – The individual or entity that has custodial control of the crane.
- *Crane User* – Arranges for the crane's presence and use at the site.
- *Engineer* – Provides engineering support and documentation for the operation.
- *General Contractor/Construction Manager* – Meets the contractual requirements for establishing and implementing acceptable safety and work performance results.

- *Lift Director* – Establishes the lift category, and reviews and implements the lift plan.
- *Lift Planner* – Develops the lift plan.
- *Rigger* – Performs rigging tasks associated with the lift.
- *Signal Person* – Directs the movements of the crane.
- *Site Safety Officer* – Enforces site safety policies.
- *Site Supervisor* – Oversees the work on the site.
- *Spotter* – Observes and reports the movement of the crane and load.
- *Transport Operator* – Operates transport equipment (dump trucks, flatbeds, barges, etc.) related to the lifting operation.

1.1.3 Potential Hazards to Personnel

Lift planning should include an evaluation of whether the lift has the potential to injure workers, bystanders, or the public. Potential hazards to consider include (but aren't limited to) the following:

- Pinch and crush points
- Personnel lifts (*Figure 1*)
- Moving or suspending loads over public areas
- Moving hazardous materials

1.1.4 Hazards Near Work Area

Lift planners should identify hazards near the work area, regardless of whether those hazards are relevant to the assigned task. Some hazards that fall into this category include the following:

- Power lines (29 *CFR* 1926.1408 defines the prohibited zones for crane operations)
- Sources of radio-frequency interference that could disrupt communications
- Electromagnetic energy sources that could induce electrical discharges
- Nearby pipes, storage tanks, and other equipment
- Structures that obscure the crane operator's view of the load and/or signal persons(s)
- Active roads and pedestrian pathways

Figure 1 Lifting personnel by crane requires special precautions.

1.1.5 Complexity of the Activity

Some lifts are not simple or straightforward, and contingency plans must be made. Planners should evaluate the following factors that can increase the complexity of the lift:

- Load instability due to configuration, shifting of contents, or location of lift points
- Special rigging requirements
- New or rarely used work practices
- Handling loads close to obstructions
- Using multiple cranes (*Figure 2*)
- Having to turn or tilt the load during the lift
- Traveling with the load

1.1.6 Commercial Impacts

Every lift has the potential to damage the load as well as nearby structures and equipment. *ASME P30.1* identifies the following factors that lift planners should consider:

- Will failure to complete the lift cause a project delay, or adversely affect access or use of the area by the public?
- Is the load extremely valuable or irreplaceable (*e.g.*, a rare or unique historical artifact, or a work of art)?
- Will replacing the load, if lost or damaged, take a significant amount of time (*Figure 3*)?
- Will damage caused by the lift to adjacent structures create an adverse commercial impact?

Figure 2 Two or more cranes involved in a lift increase the operation's complexity.

Figure 3 Damage to this nuclear reactor vessel would be a serious setback.

1.1.7 Crane and Rigging Capacities

Ensuring that the total load does not exceed the crane's load capacity at any point during the lift is a key element in lift planning. Module 21301 of this curriculum explains how to determine the gross and net capacities of a crane in a given configuration. Factors change during a lift, however, and unanticipated events can affect crane stability. Planners evaluate things that can influence crane capacity and stability, such as the following:

- Changes of the load control during operations involving demolition, suction, or friction
- Abrupt changes in the load's or crane's motion
- Changes in available line pull (*e.g.*, hoist drum effective diameter)
- Capacities of the crane's load-handling mechanisms (hydraulics, brakes, clutches, etc.)
- Accuracy of the load's weight information
- Shifting of the load center of gravity (CG) during the lift
- Shift of weight distribution between multiple cranes lifting a load
- Buoyancy effects for submerged lifts
- Out-of-vertical lifts

Less obvious but important factors to consider are the crane's work history and material condition. A manufacturer's load chart assumes the crane is in as-built condition. However, leaking hydraulic actuators or line fittings, deficient hydraulic pumps, slipping clutches and brakes, and damaged wire rope can limit a crane's capacity.

1.1.8 Environmental Conditions

Environmental conditions refer to the surroundings that can influence the lift. Environmental factors that are outside of a lift planner's control include the following:

- Wind
- Precipitation
- Lightning
- Temperature
- Conditions that can affect visibility, such as fog, wind-born dust, and the sun's position

Planning a lift for a specific time of day can minimize the effect of some natural factors, such as the sun's glare or the temperature.

1.1.9 Repetitive Lifting Operations

Examples of repetitive lifts may include tasks such as positioning construction materials (*Figure 4*), excavating fill, or performing dredging opera-

Figure 4 Filling concrete forms may require repetitive lifts.

tions. Lift planners must evaluate the effects of repetitive operations on both the equipment and the operator. *ASME P30.1* notes that planners should consider the potential for operator and construction worker fatigue and complacency. Planners should also review the crane manufacturer's recommendations for equipment duty cycles and any additional maintenance required.

1.1.10 Standard and Critical Lifts

After reviewing the results of all these evaluations, the lift director shall identify the category of the lift. There are two lift categories—standard lifts and critical lifts. If the lift director concludes that the lift does not present any undue hazards and can be completed using standard procedures, then it is classified as a standard lift. If the lift director's evaluation shows that the lift requires additional preparation, planning, or methods beyond those for a standard lift, then it is classified as a critical lift. Critical lift plans must always be written.

ASME B30.5—Mobile and Locomotive Cranes, Nonmandatory Appendix A, identifies other factors that could cause an operation to be designated a critical lift, listed as follows:

- The total load exceeds some value stipulated by company policy.
- The total load exceeds a predetermined percentage of the crane's load chart capacity (other than 75 percent).
- Hoisting personnel in a basket or platform.
- Blind lifts (any lift where the load can't be directly observed by the crane operator and/or the primary signal person).
- Any operation with two or more cranes lifting a common load. Use of tailing cranes may not require designation as a critical lift.
- Unique, costly, or irreplaceable loads.
- Demolition operations where the actual weight of the load may be in doubt (*Figure 5*).
- Especially hazardous lifts involving proximity to high voltages, lifts over people or roadways, etc.
- Other unspecified crane activities that are technically challenging.

Organizations may enforce policies that identify critical lift conditions. For example, shipyards and facilities servicing nuclear warships require that all lifts involving radioactive materials, nuclear power plant components, or weapons require a critical lift plan.

ASME P30.1 distinguishes between standard and critical lifts when describing what it requires a planner to do. For standard lift plans, it qualifies all planning actions with the term *should*, indicating that the item may be needed or recommended, depending on the circumstances, but isn't mandatory. The standard depends on the knowledge and experience of the lift director to understand the detail needed for each item.

When developing a lift plan, planners should review all the lift planning items listed in Chapters 4 (*Standard Lift Plan*) and 5 (*Critical Lift Plan*) of *ASME P30.1*, to the extent they are applicable. Chapter 5 covers the same items as Chapter 4, but describes each item in much greater detail. The standard makes it clear that in a critical lift plan, some actions are mandatory; it uses the term *shall* to identify those requirements. *ASME P30.1* also states that the critical lift planning process should address "any additional considerations identified during the planning process."

1.2.0 Developing a Lift Plan

Every lift requires a lift plan appropriate to the operation. Based on the principles found in *ASME P30.1,* this is true regardless of whether the crane is offloading bundles of 2 × 4 studs from a flatbed

Figure 5 Demolition lifting operations may be designated critical lifts.

truck or placing a 278-metric ton reactor vessel at a power plant. The overall approach is the same for either operation; only the scope and level of detail of the planning differs. Standard lift plans may be verbal or written. Critical lift plans shall be in written form.

The following sections allow you to assume the role of lift planner, documenting the information necessary to produce a lift plan based on a provided scenario. Then, in the role of the lift director, you will categorize the lift as a standard or critical lift. The scenario for this exercise is described as follows:

This site construction plan calls for the installation of an air-conditioning unit on the third floor roof of a building. *Appendix A* provides an architectural elevation and roof plan of the building, showing the unit's installed location and orientation. *Appendix B* supplies technical details of the unit related to the lift, including shipping weight, center of gravity, and minimum rigging requirements. *Appendix C* is the load chart for the Grove TM1500 crane selected for the lift. Assume that the site can provide any necessary materials, workers, and other equipment for the lift.

1.2.1 The Lift Data Sheet

Planning even a simple lift can be a challenging task. Many factors described in *ASME P30.1,* Chapter 2 are interdependent. Breaking the process into distinct steps in a logical order can help to provide clarity. Completing a lift data sheet (LDS) is the preferred method for ensuring that all applicable lift planning items have been considered.

An LDS is a summary of the essential technical details relating to the lift. *ASME P30.1, Nonmandatory Appendix A—Example Lift Plan*, includes a generic LDS form that can be the basis for users to develop their own customized forms, which the standard encourages. A brief internet search for the term *crane lift planning form* demonstrates the variety such forms can take. The form can contain the following informational sections, as well as others that companies, lift directors, or lift planners may consider necessary:

1. Lift description and other identifying information
2. Identification of the crane and its configuration
3. Identification and description of the load and its rigging requirements
4. Computation of crane capacities
5. Reference attachments and supporting documentation
6. Notes
7. Review and approvals

Appendix D is a blank LDS form based on this format. Refer to this form as the development of the lift plan proceeds through the following sections.

1.2.2 The Crane Configuration

After filling in the project information at the head of the LDS, the first section relates to the crane and its configuration. Enter the crane's identifying information, referring to the crane's load chart and operator's manuals. Before you can determine a crane configuration, you need to know the load's weight and the geometry of the lift in relation to the available setup areas.

The building's roof plan (*Appendix A*) shows that the condensing unit is located toward the south end of the building. This means that the crane should be positioned at that end of the building, and on a line perpendicular to the wall through the load's landing site to minimize the load radius (*Figure 6*). Minimizing the load radius maximizes capacity and crane stability. Before proceeding, check with the project supervisor or schedule to determine if placing a crane at that location on the scheduled day is acceptable.

Consult the crane's range diagram (*Appendix C*, page 1 of 6). Note that the roof edge is 34'-6" (10.5 m) above the ground. The center of the condensing unit will be 47'-8" (14.5 m) from the roof edge. It helps to sketch this information onto a sheet of graph paper or use tracing paper taped over the

crane's range diagram. *Figure 7* shows the building elevation view superimposed on the range diagram.

Assuming that the crane will conduct the lift with the main boom over the rear, as shown in the range diagram, determine the position of the crane that results in the shortest load operating radius and also provides sufficient clearance to the building for the elevated boom. This will give you the boom angle and boom length for the lift. Note that these are tentative values; the crane's capacity in this configuration must also be determined.

The following working assumptions can be made from analyzing the situation and the range diagram with the building's outline and load location placed on it to scale:

- The crane will be on fully-extended outriggers, with its center of rotation about 25 feet (7.6 m) from the building wall. The crane body will be on a line perpendicular to the wall passing through the condenser's location. This arrangement allows the planner to rely on the range diagram directly, without adjustment. Oblique (angular) positions to the load in relation to the building require calculations using dimensions from the architectural plan and elevation views, or they must be physically measured on site.
- The range diagram shows that the minimum boom angle permitted in this location is about 40 degrees. A lower angle with the crane in the described position would place the boom too close to the edge of the roof. The load radius is 73 ft (25 ft + 48 ft = 73 ft), so a minimum boom length of 106 ft (32.3 m) at a 40-degree boom angle can be used to place the unit. The range diagram also shows that the load can be set with the main boom fully extended to 141 ft (43 m) at a boom angle of about 55 degrees.
- The power-pinned boom fly is retracted.

Consult the crane's load chart to determine whether the trial combination of load radius and boom length is within the capacity of the crane. If there is plenty of excess capacity, then you, as the lift planner, may confidently proceed with filling in the crane data on the LDS. If the load's weight appears to be a significant percentage of the crane's capacity (before weight deductions) then you must evaluate other possible configurations. Options may include lifting from a jib/boom extension, or simply trying the longer boom length and boom angle noted above.

When planning reveals the operation will be at the limit of the crane's capacity, finding an improved configuration might require several

NCCER – *Advanced Rigger*

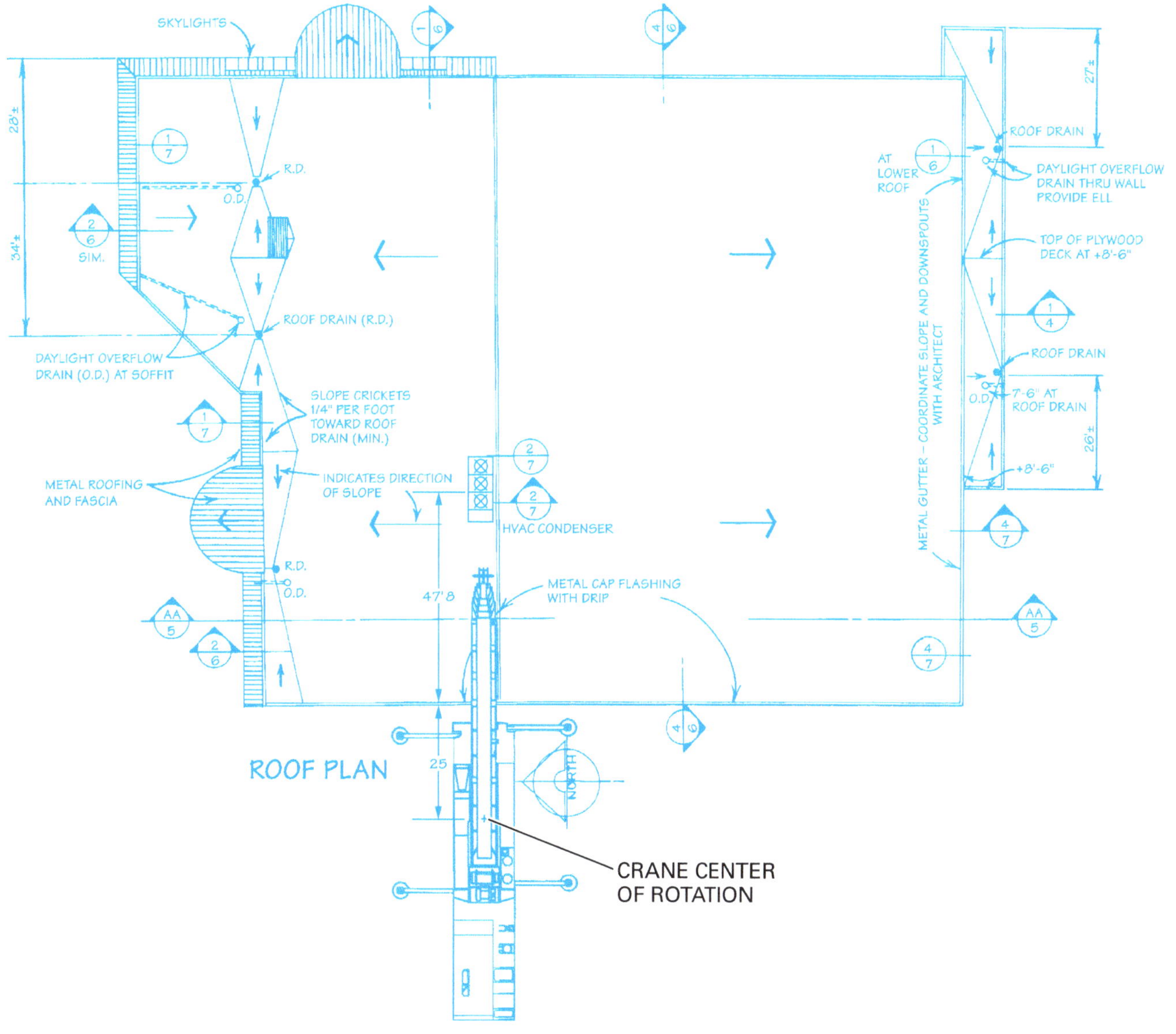

Figure 6 Plan view showing the position of the crane for placing the HVAC condenser.

Lift Planning Software

It shouldn't be surprising that there are apps available that can assist with lift planning. These apps run on desktop and laptop systems, tablets, and smartphones. Some companies offer free basic planning programs, such as 3D Lift Plan™, that run in a browser environment along with their full-featured programs. These applications allow planners to visualize the site structures, crane, and load in 3D. Users can see the geometry of the lift and the path the load will travel.

The main limitation of these apps is that they don't provide access to a specific crane's load chart to evaluate capacity variations during the lift. Some crane manufacturers, such as Manitowoc, have resolved this limitation by offering planning programs such as CraniMAX® that include an update-able crane database containing specific crane load charts.

A brief search of the internet using the search term *lift planning software* will reveal numerous low- to high-end programs offered by software companies, crane manufacturers, and crane dealers. However, as with all computer programs, the output is only as good as the input. Lift planners are still required to verify the results.

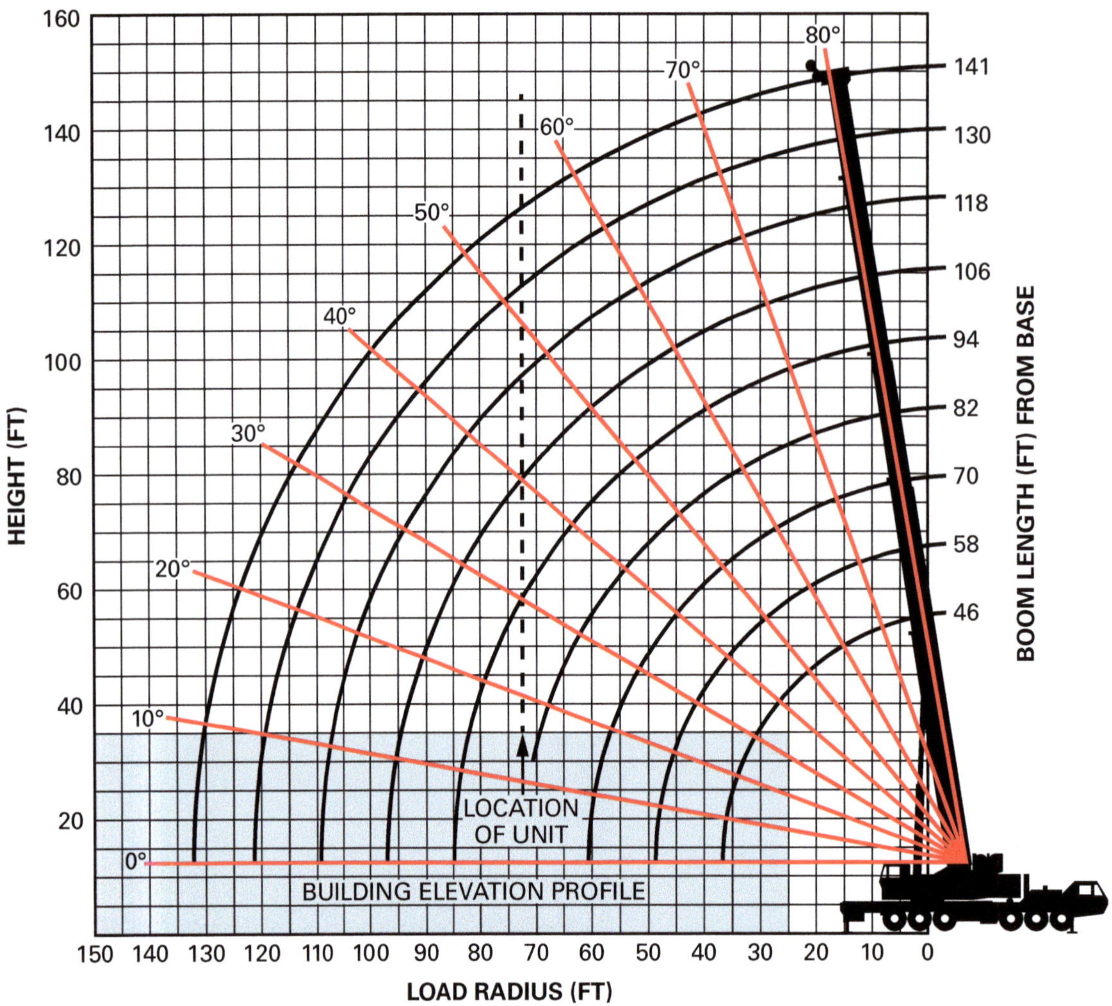

Figure 7 Elevation view of building on the crane's range diagram.

attempts. You may have to modify the crane's configuration more than once, also taking into account the suspended weights of hoist components, rigging, load, and other deductions. You may have to relocate the crane, or even choose another crane altogether. (Refer to Module 21301, which describes in detail how to perform these calculations.)

On the load chart, the load radius of 73 ft (22.3 m) falls between the listed 70 ft load radius and 80 ft load radius (*Figure 8*). Select the 80 ft load radius, because it will yield lower, more conservative crane capacities. Read across the row looking for a capacity with the boom at an angle roughly between 40 degrees and 55 degrees.

You will see that there are several options. Referring to the load chart (*Figure 9*), with the boom at 106 feet (32.3 m) long, the boom angle is only 39 degrees, which places it too close to the building's roof. At boom lengths of 118 ft (36 m) and 130 ft (39.6 m), higher boom angles are permissible. The crane's gross capacity at both

boom lengths is 18,850 lb (8,550 kg). The capacity is 18,700 lb (8,482 kg) at the maximum main boom length of 141 ft (43 m), but the boom angle must be greater than 55 degrees due to structural limitations. The crane can't reach the condenser location at this angle. The 18,850 lb capacity is more than six times greater than the anticipated unit weight of 2,797 lb (1,269 kg) shown in *Appendix B*. For this lift, the load chart shows that the 130 ft boom length will provide the best stability and the greatest clearance to the building when swinging the boom.

> **CAUTION**
>
> Operators must not interpolate crane load chart capacities. Whenever the load radius, boom length, or boom angle falls between listed values on a load chart, the operator must select the adjacent value that yields a lower capacity. This practice provides an additional margin of safety, ensuring the load does not exceed the crane's capacity.

NCCER – *Advanced Rigger*

RATED LIFTING CAPACITIES IN POUNDS
46 FT – 173 FT BOOM ON OUTRIGGERS – OVER REAR

Radius in Feet	#0001 Main Boom Length in Feet (Power Pinned Fly Retracted)									#0002 Power Pin F'v Ext. & 141 ft
	46	58	70	82	94	106	118	130	141	172
10	300,000 (74.5)									
12	280,000 (72)	143,500 (76)	142,000 (79)							
15	235,000 (67.5)	143,500 (76)	141,500 (72)	130,000 (79.5)						
…					(57)	…,050 (61.5)	40,700 (65)	38,250 (68.5)	35,750 (71)	32,… (75.5)
60			36,299 (22.5)	36,200 (38.5)	36,200 (48.5)	34,300 (55)	33,600 (58.5)	30,750 (63.5)	28,500 (66.5)	26,350 (72)
70				26,050 (25)	26,050 (39.5)	26,050 (47.5)	26,050 (53)	24,750 (56)	23,100 (61.5)	22,000 (68.5)
80					18,850 (27)	18,850 (39)	18,850 (46.5)	18,850 (52.5)	18,700 (56.6)	18,500 (64.5)
90						13,500 (28)	13,500 (38.5)	13,500 (46.5)	13,500 (51.5)	15,250 (60.5)
100							9,390 (29)	9,390 (39)	9,390 (45.5)	17,500 (56.5)
110							6,080 (12.5)	6,030 (30.5)	6,080 (39)	10,100 (52)
120								3,390 (17.5)	3,390 (31)	7,530 (47.5)
130									1,150 (19.5)	5,390 (47.5)
140										3,610 (36.5)
150										2,100 (30)

Figure 8 Evaluating the crane's available boom configurations for the lift.

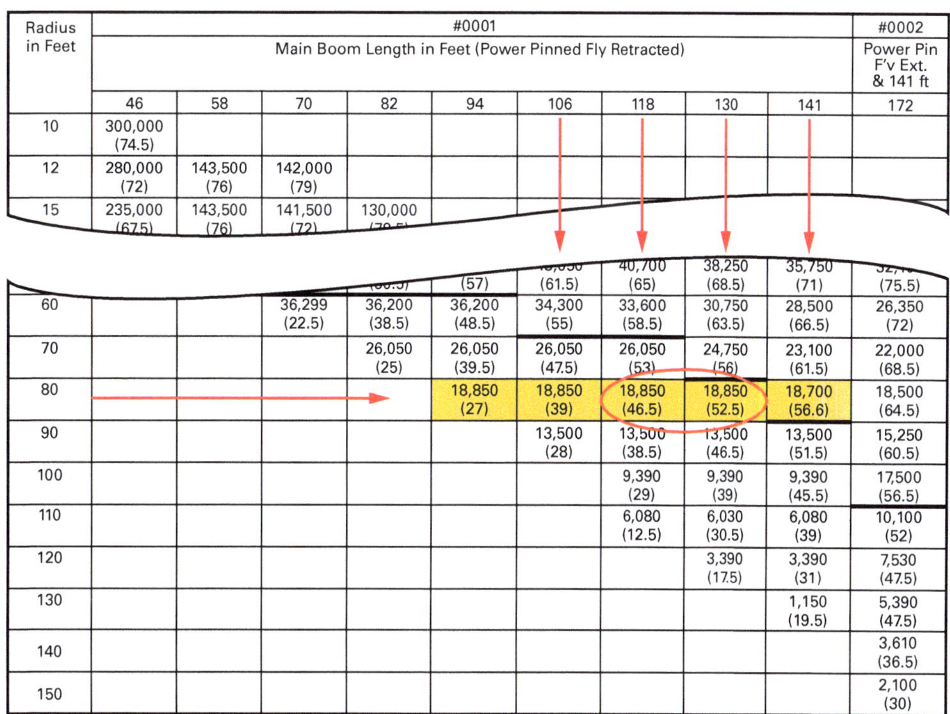

RATED LIFTING CAPACITIES IN POUNDS
46 FT – 173 FT BOOM ON OUTRIGGERS – OVER REAR

Radius in Feet	#0001 Main Boom Length in Feet (Power Pinned Fly Retracted)									#0002 Power Pin F'v Ext. & 141 ft
	46	58	70	82	94	106	118	130	141	172
10	300,000 (74.5)									
12	280,000 (72)	143,500 (76)	142,000 (79)							
15	235,000 (67.5)	143,500 (76)	141,500 (72)	130,000 (79.5)						
…					(57)	…,050 (61.5)	40,700 (65)	38,250 (68.5)	35,750 (71)	32,… (75.5)
60			36,299 (22.5)	36,200 (38.5)	36,200 (48.5)	34,300 (55)	33,600 (58.5)	30,750 (63.5)	28,500 (66.5)	26,350 (72)
70				26,050 (25)	26,050 (39.5)	26,050 (47.5)	26,050 (53)	24,750 (56)	23,100 (61.5)	22,000 (68.5)
80					18,850 (27)	18,850 (39)	18,850 (46.5)	18,850 (52.5)	18,700 (56.6)	18,500 (64.5)
90						13,500 (28)	13,500 (38.5)	13,500 (46.5)	13,500 (51.5)	15,250 (60.5)
100							9,390 (29)	9,390 (39)	9,390 (45.5)	17,500 (56.5)
110							6,080 (12.5)	6,030 (30.5)	6,080 (39)	10,100 (52)
120								3,390 (17.5)	3,390 (31)	7,530 (47.5)
130									1,150 (19.5)	5,390 (47.5)
140										3,610 (36.5)
150										2,100 (30)

Figure 9 Crane capacities at the available boom lengths.

Fill in the applicable blanks in the crane section of the LDS with the information that you know or can assume at this point. More data will become available later in the planning that may alter some of this information.

1.2.3 The Load and Rigging

The next section of the LDS pertains to the load and its characteristics, the required rigging, and the hoist components needed. The HVAC unit data sheet (*Appendix B*) indicates the shipping weight is 2,797 lb. You should use this value as its hoist weight, even if some shipping or packing materials are removed before the lift.

The data sheet also identifies the manufacturer's minimum lifting and rigging requirements (*Figure 10*). The condenser is factory mounted on a structural steel skid which includes four lifting lugs. The manufacturer recommends suspending the load from four slings shackled to the skid and using a spreader system to stabilize the slings. The manufacturer also suggests attaching all four slings directly to the lifting hook.

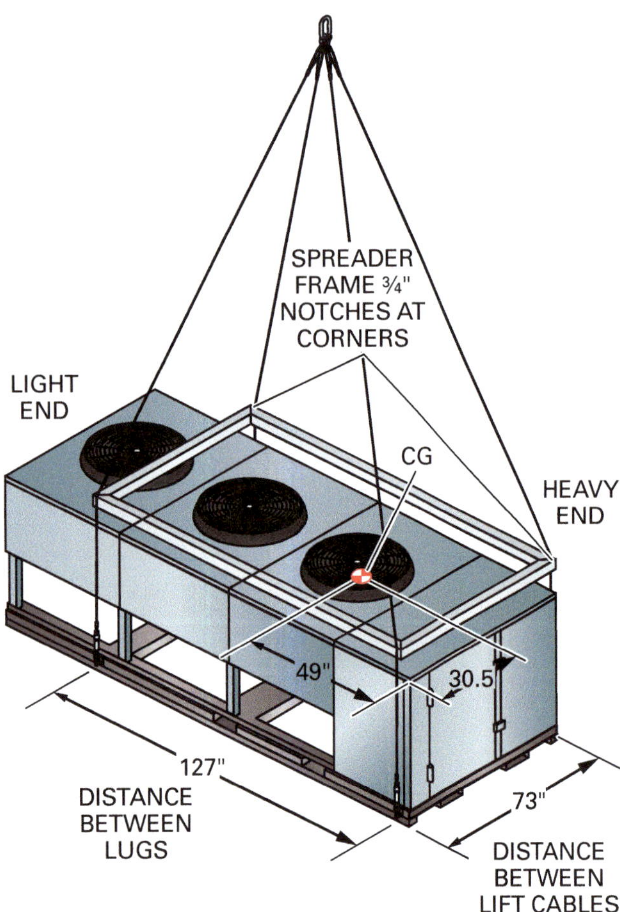

Figure 10 Rigging diagram for the HVAC unit.

Riggers have determined that the condenser will require the following rigging components:

- Four (4) 2,500 lb shackles (0.2 lb each)
- One (1) hoist shackle (2.3 lb)
- Four (4) 12-ft wire rope slings (3.9 lb each)
- One (1) 2 × 4 wooden spreader frame, 74" × 121" (100 lb)

Use this information to fill in the LDS load and rigging weight section.

Part of the planning for this lift includes determining what other hoisting hardware the crane will have to carry on the day of the lift, and the need for any alteration of the crane's configuration. Crane owners don't change the hoisting hardware configuration unless needed, so you may be able to obtain the information ahead of time for planning purposes. For this lift, assume the crane is equipped with the following hoisting components identified in the load chart:

- 150-ton, 8-sheave block
- Auxiliary boom head installed
- 10-ton headache ball (suspending the load)

Enter the weights of all applicable hoist items, then find the sum of the weights in the form to obtain the total weight acting on the crane boom:

```
      2,797 (HVAC condenser weight)
        119 (rigging)
      5,254 (8-sheave block)
        261 (auxiliary boom head)
  +     560 (10-ton headache ball)
      8,991 total lifted weight
```

1.2.4 Crane Capacity Calculations

Accurately completing the crane capacity section of the LDS is essential to a successful lift plan. The lift director will use the capacity information as a factor to determine the lift category—standard or critical. The capacity section provides three columns, permitting calculations at different points in the lift. The planner may need these for variations in the load radius that occurs during the lift.

Before developing these estimates, a planner needs to know where the load will be staged for lifting. If the project manager doesn't know, suggest a spot that will make planning easier. One of the columns should reflect the capacity at the staging area load radius. Another column should show the capacity at the landing site. For each column of the LDS, follow these steps to determine the percent capacities:

Step 1 Identify at the column head the limiting condition for which you made calculations.

Step 2 On the next row, enter the total lifted weight calculated in the previous section.

Step 3 Enter the actual load radius for the conditions described in the column.

Step 4 Enter the load chart load radius used to determine gross capacity. This assures the lift director that you applied the proper lower capacity for a given load radius.

Step 5 Enter the load chart capacity for the configuration.

Step 6 Divide the total weight lifted by the load chart capacity in that column. Enter this value as a percentage (%).

For the condenser destination on the roof:

$$\% \text{ of chart capacity} = \frac{\text{total lifted weight}}{\text{load chart capacity}} \times 100$$

$$\% \text{ of chart capacity} = \frac{8,991 \text{ lb}}{18,850 \text{ lb}} \times 100$$

$$\% \text{ of chart capacity} = 0.477 \times 100 = \textbf{47.70\%}$$

Enter the largest percentage of load chart capacity you calculated in this section in the next line; the text formatting of this line draws reviewers' attention to it.

The following line evaluates the capacity of the reeved hoist line in relation to the total weight. Note that the lift will use the single-part auxiliary hoist with the 10-ton headache ball hook. Consult the crane specifications to determine the minimum breaking strength (MBS) of the hoist cable. For this crane, the manufacturer recommends $\frac{3}{4}$" IWRC rope for the auxiliary hoist, which has a MBS of 51,800 lb, per the vendors specifications.

For the total load, you need to add up only those weights lifted by the hoist cable:

2,797 (HVAC condenser weight)
119 (rigging)
+   560 (10-ton headache ball)
3,476 total suspended weight

Enter this result in the Total Suspended Load blank. Enter the hoist line MBS in the Reeved Capacity blank. Divide the suspended weight by the line MBS and multiply by 100%:

$$\text{max \% reeved capacity} =$$
$$\frac{\text{total suspended weight}}{\text{hoist-line MBS}} \times 100 =$$

$$\frac{3,476 \text{ lb}}{51,800 \text{ lb}} \times 100 =$$

$$0.067 \times 100 = \textbf{6.70\%}$$

Enter the result in the Maximum % Reeved Capacity blank on the LDS (*Figure 11*). Industry standards specify that the ratio of maximum load to ultimate strength of a lifting system should be less than 1:5, or 20 percent.

1.2.5 Load and Crane Considerations

For the HVAC unit lift, the crane capacity and rigging configuration appear to be well within permitted limits. However, *Appendix B* shows that the condenser's CG is not exactly in the center of the lift attachment pattern. It is located closer to one end of the unit. There is some potential for the out-of-balance load to shift or become unstable during the lift. When a load's CG is off-center, the rigging supervisor may require lifting the load with an adjustable equalizer beam. Depending on the design of the beam, riggers can position the beam's lifting lug(s) so that the load's CG is directly below the load block.

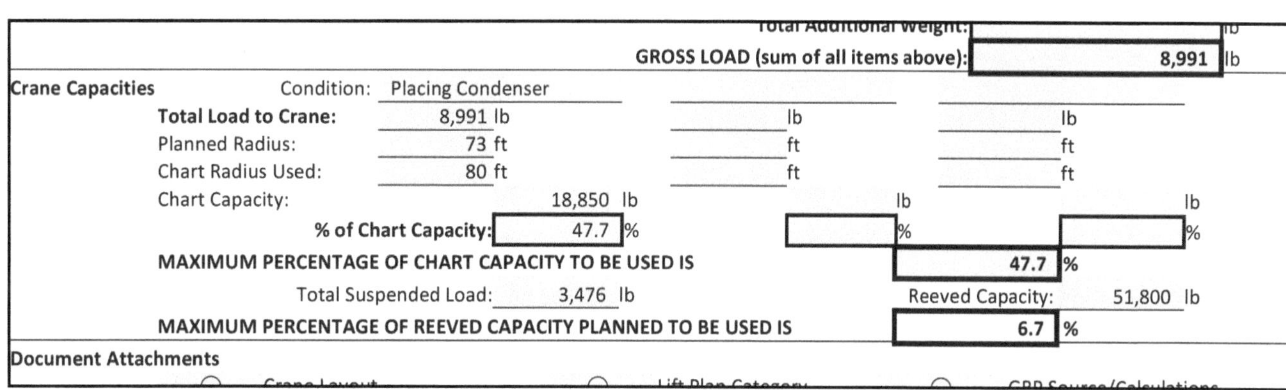

Crane Capacities						
			Total Additional Weight:			lb
			GROSS LOAD (sum of all items above):		**8,991**	lb
Condition:	Placing Condenser					
Total Load to Crane:	8,991 lb		lb		lb	
Planned Radius:	73 ft		ft		ft	
Chart Radius Used:	80 ft		ft		ft	
Chart Capacity:	18,850 lb		lb		lb	
% of Chart Capacity:	47.7 %		%		%	
MAXIMUM PERCENTAGE OF CHART CAPACITY TO BE USED IS				47.7 %		
Total Suspended Load:	3,476 lb		Reeved Capacity:	51,800 lb		
MAXIMUM PERCENTAGE OF REEVED CAPACITY PLANNED TO BE USED IS				6.7 %		
Document Attachments						

Figure 11 Completing the Crane Capacities section of the LDS.

Note in the rigging figure of *Appendix B* that the lifting lugs at the heavy end of the condenser are 49" lengthwise from the condenser's CG. The distance between the condenser lifting lugs along the side of the unit is 127". Therefore, the distance between the unit's CG and the light-end lugs is:

$$127.00" - 49.00" = \mathbf{78.00"}$$

If using an equalizer beam, the riggers position the heavy-end suspension lugs at 49" from the lifting eye, and 78" from the lifting eye for the light-end lugs (*Figure 12*). Note that the front-to-back position of the condenser's CG is also off center, but the distance is acceptable. For this lift, the condenser manufacturer has determined that a single-point lift will provide sufficient stability to place the condenser, so an equalizer beam is not required.

A single crane can place the HVAC condenser on the roof of the building as described in this scenario. However, there are situations when, either due to the excessive weight or the awkward location of the landing site, multiple cranes must handle the load (*Figure 13*). This curriculum provides information on multi-crane lifts and the relevant load calculations in Module 21303.

The lift planner may consider lifting a load with two cranes attached to each end of an equalizer beam. For loads having CGs near their geometric centers, the load lifted by each crane is nearly the same. Determining crane capacities for such lifts is fairly simple.

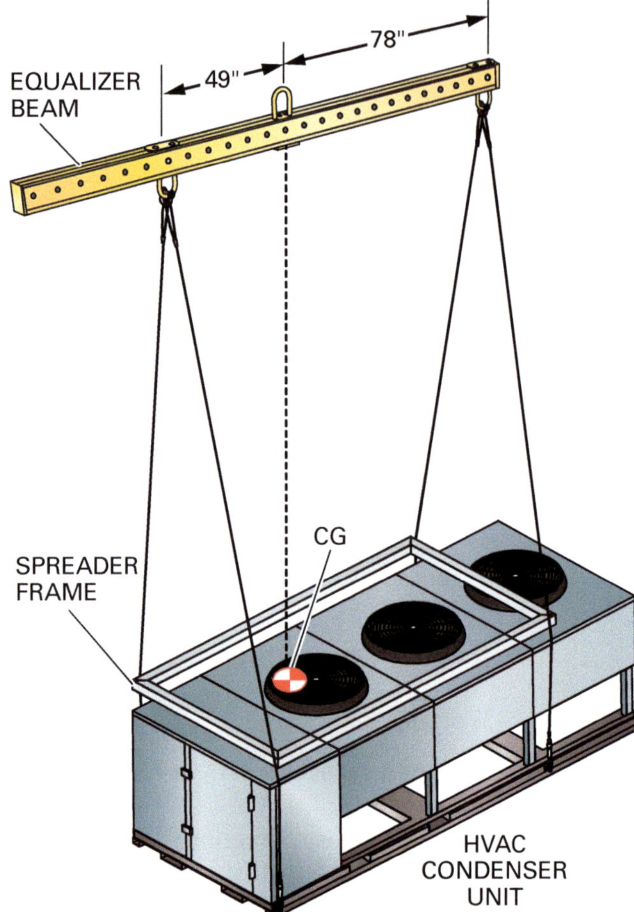

Figure 12 Positions of the adjustable suspension lugs on an equalizer lifting beam.

Figure 13 When positioning a load becomes complicated, two or more cranes may be required.

NCCER – *Advanced Rigger*

For a load with the CG well off-center, the calculations are more complicated. The crane hoisting the heavy end of the load will use a higher percentage of its gross capacity than an identical crane hoisting the lighter end, everything else being equal. If the two cranes have different gross capacities (which is common), the calculations become even more difficult. The lift planner must carefully consider the movement of the load and the capacities of the cranes at every moment and position during the lift. Planning and coordinating dual-crane lifts can be very complex. For this reason, multi-crane lifts are always in the critical lift category.

1.2.6 Reference Documents and Attachments

The Documents and Attachments section of the LDS is a checklist of typical resources that a lift planner may need to consult or include in the lift plan for supervisory review. The HVAC unit lift plan should probably include, at a minimum, the following information or documents:

- Crane layout and configuration
- Crane load chart extract
- Load/crane clearances
- Drawing of load
- Load weight and CG information
- Rigging arrangement
- Crane foundation evaluation
- Ground bearing pressure (GBP) data/calculations
- Job hazard analysis (JHA)
- Weather forecast

All crane-related information can be appended to the plan by including the crane's load chart or a certified copy of the relevant sections. The attachment should provide, if possible, a sketch of the crane setup for the lift, and/or include a description of the crane and lifting accessories. It should also include the crane's range diagram and tables of deducted weights. The lift director or site policies may require that each photocopied page bear a signature certifying that it is a true copy of the load chart.

The minimum building clearance for the boom can be determined from the range diagram and building elevation plan superimposed on it to scale, as shown in *Figure 14*. Policies may require a more formal estimate performed by the project engineer, who will produce a scale drawing illustrating the minimum lift clearances. The planner should add a note to the LDS stating that, before swinging the load over the roof, the operator must ensure the boom angle equals or exceeds the load chart angle specified in the planning.

Planners should reference the HVAC unit manufacturer's specifications, which will illustrate the unit in three views, as in *Appendix B*. The plans will provide the unit's dimensions, lift points, center of gravity, and the recommended rigging. If policies require a paper copy of this information, include only the pages of the manual relevant to the lift.

The rigging arrangement should be at least an accurate sketch of the intended rigging accessories, slings, and attachments. For complex lifts, the project engineer or the load's manufacturer may provide a scale drawing of the rigging and an itemized list with rigging weights. Note that the specified weights of equipment are typically dry weights. The addition of any fluids to the equipment prior to the lift must also be considered. For example, a power transformer is quite heavy alone, but a significant amount of weight is added if it is lifted into position after it is filled with oil.

Include attachments for the crane foundation, matting, and GBP if the project engineer determines these items are needed (*Figure 15*). For this lift, the ground that will support the crane is undisturbed, firm, and level. The project engineer has previously evaluated the GBP for earlier lifts at this site and has determined that ground support is adequate. The planner should include a memo to this effect in the lift plan package.

Before finalizing the plan, consult with the site safety officer for a job hazard analysis (JHA). Discuss the crane communications that will be used, the number of signal persons required, access and traffic controls needed, PPE, and any other site-specific policies that apply. Summarize this information in an attachment and include any operational notes on the LDS. Site or employer policies may require the safety officer to sign the document. *ASME P30.1, Nonmandatory Appendix A—Example Lift Plan*, includes an example of a pre-lift safety checklist used to summarize this information. The lift director can also use the form to document the safety portion of a pre-lift briefing.

An efficient lift planner always checks the weather forecast for the lift date to identify any potential problems. If the forecast is for windy conditions, thunderstorms, or heavy rains, this is clearly noted in the lift plan. Postponements for weather are relatively common. Planning should include go/no-go criteria for any contingencies that may arise.

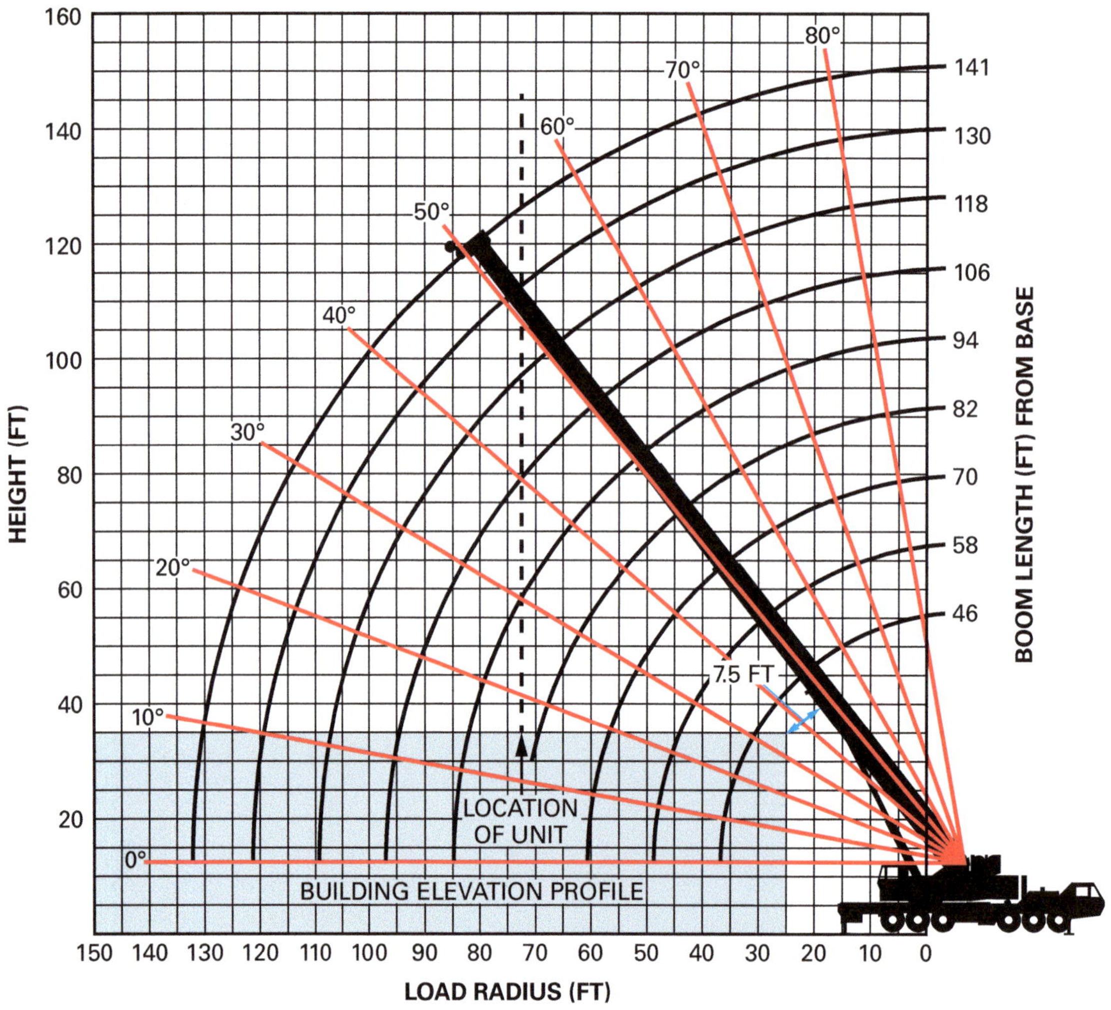

Figure 14 Determining building clearance from the crane range diagram.

Figure 15 Inadequate ground bearing pressure (GBP) information can lead to crane tipping incidents.

1.2.7 Notes, Review, and Approvals

Identify any information that reviewers should be aware of in the Notes section of the LDS. In addition to those already discussed, considerations that might be listed in this section include the following:

- Anything that might potentially limit crane capacity or re-categorize the lift
- People and equipment (*e.g.*, taglines) required for controlling the load during the lift
- Crane communication details and the signal persons/spotters required
- Establishing access and traffic controls during the lift
- Referencing emergency action plans and contact information in the event of accidents

When the planner completes the lift plan package, the next person to receive it may be the project engineer, site safety officer, or the lift director. The lift director reviews the lift plan and identifies the category of the lift.

Assume that you are the lift director for the HVAC unit lift. When categorizing the lift of the condenser, consider and answer the following questions:

- Can a single crane conduct the lift?—**Yes**
- Is the total weight lifted by the crane less than 75% of its determined capacity?—**Yes**

> **NOTE**
>
> While 29*CFR* 1926.751 classifies any lift greater than 75 percent of load chart capacity for steel erection as a critical lift, this technically applies only to steel erection. Many organizations, however, apply the same percentage of capacity to all lifts when determining if the lift is a critical lift. Organizations may establish both higher or lower percentages of capacity as the threshold for a critical lift when it is not related to steel erection, but for steel erection, the threshold can be lower than 75 percent, but not higher.

- Is the unit or its contents inherently hazardous, contain materials immediately dangerous to life or health, or extremely valuable, hard to replace, or irreplaceable?—**No, but any refrigerant it may contain is controlled by US Environmental Protection Agency (EPA) Clean Air Act, and must not be vented to atmosphere. An unintentional release, depending on the volume, may require reporting to the EPA.**
- Is the load's CG fixed and is the load considered stable?—**Yes**
- Does the lift present any unusual hazards to the public, utilities, or infrastructure?—**No**
- Are there any unique hazards to the load or crane?—**Only the building itself; accounted for in the plan**
- Are there any site or other hazards that could reclassify the lift?—**None identified**

Based on this evaluation, classifying it as a standard lift is justified. Lift directors understand that this is a tentative classification that can change at any time if an unanticipated factor is encountered.

1.2.8 Critical Lift Plans

ASME P30.1 requires that lift plans for critical lifts be in a written form. Due to the complexities or potential hazards involved, critical lift plans can be large, detailed packages, addressing all applicable sections discussed in Chapter 5 of *ASME P30.1*. Since there is no set format for lift procedures, critical lift plans differ in scope and content with each lift, by company policy, and from one work site to another.

Development of a critical lift plan follows a process similar to that for a standard lift. However, the necessary details usually require the input of multiple parties. *Appendix E* provides an example of a critical lift planning form.

In addition to sections devoted to definitions, training, and applicable regulations, these plans often include a detailed lift procedure. The procedure guides workers step-by-step through preparation for the lift, staging the load, and installing the load's rigging. For high-risk lifts involving dangerous materials or high-value loads, the procedure may require signatures for completion of each step. A critical lift may also require a test lift using a dummy weight. If the lift involves precision assembly or securement of the load as it is placed, the procedure may specify the need for other crafts and quality assurance inspections.

1.3.0 Implementing a Lift Plan

If the planning is thorough, putting a lift plan into action is straightforward, if not routine. Project managers work with their team to set the plan in motion. Efficient implementation of a lift plan requires frequent communication among the participants. A post-lift review is often done to identify issues to consider in future lifts.

1.3.1 Pre-Lift Review/Meeting

All participants need to review the lift plan. This review may be for final approval, or part of a progression towards a final plan. The review may identify conditions or requirements that are different than those originally considered. Reviewers inform the lift planner in such cases, especially if the information could result in re-classifying a lift as critical. Even if project supervision has reviewed and approved a lift plan prior to the scheduled activity, key participants should review the plan again before conducting a pre-lift meeting. *ASME P30.1* notes that the lift director should determine how often to review the lift plan during a series of lifts. It isn't usually necessary to review the lift plan between each lift.

The complexity and category of a lift should dictate the scope and formality of a pre-lift meeting. For simple jobs or repetitive lifts, a simple face-to-face meeting at the work site involving the crane operator, signal person, rigger, and the supervisor responsible for the load may be sufficient. For the initial meeting of a complicated and/or critical lift, gathering in an office may be necessary (*Figure 16*). Audiovisual services can make such a meeting more effective. Participants should include the site supervisor or representative, site safety officer, site engineer, lift director, rigging supervisor, craft supervisor responsible for the load (if applicable), the crane operator, all signal persons, and any other individuals that will provide site services or traffic control.

The lift director should conduct the pre-lift meeting. Items to review may include, but are not limited to, the following:

- Main sections of the LDS, as applicable
- Rigging requirements
- Site conditions and weather
- Coordination of services
- Primary and backup crane communications
- Site policies pertaining to the lift
- Contingency plans
- Job hazard analysis, if appropriate

Critical lift meetings will usually involve significantly more detail. Policies may require completion of an JHA checklist and review at the meeting, which all participants must sign as part of a formal safety briefing.

The lift director must resolve any concerns raised at the meeting before lifting begins. It is possible that a concern could result in re-categorizing the lift, which would require additional planning or preparations. Only the designated lift director has the authority to proceed with executing the lift plan.

Figure 16 A formal pre-lift meeting is necessary for complicated or critical lifts.

1.3.2 Executing the Lift

Before a lift may commence, all preparations pertaining to the lift must be complete, including the following:

- The load must be in its designated staging location.
- The rigging must be tested, inspected, and installed per the lift plan.
- The crane must be ready in the location and configuration specified by the lift plan, including any certifications, testing, and inspections.
- All designated personnel are on location with required PPE and equipment.
- The crane operator and signal person(s) have established and tested primary and backup crane communications.
- Vehicular and personnel traffic controls are in place.
- Weather and site conditions support the lifting operations.

The lift director must determine that no condition exists that would prevent the lift from proceeding as planned. If the team encounters something unexpected (*Figure 17*), the lift director must stop the lift and evaluate the situation. The load must be placed in a safe condition during the wait, which usually means lowering it to the ground.

Any participant can issue a Stop command during the lift. *Section 5-4.3* of *ASME P30.1* lists several actions the lift director should consider if the operation deviates from the plan. If a change to the plan is necessary, the lift director should communicate the change to all participants before continuing. Only the lift director can restart a lift after it has been stopped.

1.3.3 Post-Lift Review

Conducting a post-lift review is always a good management practice. *ASME P30.1* suggests that lift directors hold reviews after both standard and critical lifts. These activities provide the opportunity to learn something new from the lifting operation. Many organizations refer to this as "lessons learned".

To be truly effective, lift directors should schedule a review as soon as possible after the operation. Participants in the review should identify any problems in the development, planning, and execution of the lift. They should also suggest improvements that could apply to future lifts or site operations in general. The lift director will then communicate any action items to the responsible parties for consideration and correction, as applicable.

Figure 17 At the onset of dangerous weather, stop the lifting operation.

Additional Resources

Planning for Load Handling Activities, ASME P30.1. New York: The American Society of Mechanical Engineers.

"Load Charts for Use With Written Examinations—Grove TM1500." Manitowoc Crane Group and NCCER. **www.nccer.org/docs/default-source/Load-Charts/grovetm1500.pdf?sfvrsn=853e0e4f_02**9

CFR 1926, Subpart CC,*Cranes and Derricks in Construction*. **www.ecfr.gov**

1.0.0 Section Review

1. The most important purpose of lift-planning standards is to _____.

    a. protect the careers of all lift participants
    b. protect lives and prevent injuries
    c. prosecute violations of crane safety
    d. provide guidance to lawmakers to create and enforce crane-related safety laws

2. The first step in determining the crane configuration for a particular lifting operation is _____.

    a. determining the minimum single-line pull
    b. calculating the number of parts of hoist line
    c. finding the crane's gross capacity from the load chart
    d. consulting the crane's range diagram

3. If a lift plan reviewer identifies a condition or requirement that differs from what the planner considered, the reviewer should _____.

    a. make a pen-and-ink change to the plan, then approve it
    b. request a second opinion from the supervisor
    c. inform the lift planner of the situation
    d. assume that the next reviewer will correct the problem

SUMMARY

A lift plan is essential to the success of a lifting operation. It is the framework that brings together and organizes all aspects of mobile crane operations presented in this curriculum. Even if safety were not the most important factor, executing a lift without proper planning would inevitably result in false starts, work delays, damage to the load or other structures, and wasted time, effort, and money.

Only qualified individuals with extensive experience in crane and construction site operations should conduct and lead lift planning. Lift planners must be able to intelligently consult with specialists in the various crane, engineering, and construction professions to put together a safe, effective plan.

All participants must know their role in implementing the plan and be qualified to do so. Lift directors must be able to communicate the plan requirements to all team members, continually monitor the operation, and be open to lessons learned after its completion.

1. The person responsible for developing the lift plan is the _____.

   a. crane operator
   b. site supervisor
   c. lift director
   d. lift planner

2. Planners would consider any of the following to be a complex crane lift *except* _____.

   a. a single crane lifting materials from the ground to the top of a structure
   b. a lift involving two or more cranes
   c. having to turn a or tilt the load during the lift
   d. lifting an unstable load

3. Planning a lift for a specific time of day could minimize the adverse effects of _____.

   a. ground bearing pressure
   b. load buoyancy
   c. the sun's glare
   d. rain or lightning

4. The term that distinguishes many of the requirements of critical lift planning from standard lift planning is _____.

   a. *may*
   b. *shall*
   c. *can*
   d. *might*

5. During lift planning, positioning the crane where the load radius is minimized _____.

   a. places unnecessary stress on the outriggers
   b. creates the least stable situation
   c. improves stability but sharply reduces capacity
   d. maximizes capacity and crane stability

6. The best choice for stabilizing an off-balance load before lifting it is to _____.

   a. add weights to the light end of the load to balance it
   b. hook a tag line to each end so the riggers can manipulate it
   c. use an adjustable equalizer beam
   d. use a second crane

7. The individual responsible for categorizing the lift operation is the _____.

   a. site superintendent
   b. lift director
   c. lift planner
   d. crane operator

8. A concern raised during a pre-lift meeting could result in _____.

   a. re-categorizing the lift
   b. changing the lift planner
   c. revising the crane's load chart
   d. cancellation of the lift

9. Which factor, by itself, would *not* normally delay a scheduled lift?

   a. Discovering the load rigging does not have inspection tags.
   b. The load weight is greater than planned.
   c. Clouds blocking the sun.
   d. The crane's configuration is different from the lift plan.

10. Participants should conduct post-lift reviews to _____.

    a. establish blame for any mistakes that occurred during the lift
    b. obtain lessons learned for planning future operations
    c. approve the lift plan
    d. critique the work of the lift planner

Ground bearing pressure (GBP): The maximum force per unit area (pressure), measured in pounds per square foot, that can be exerted on the ground's surface without causing deformation of the surface. Can only be determined by testing.

Lift data sheet (LDS): The essential page or pages of a written lift plan that organize and summarize the important information required to develop a lift plan.

Building Architectural Plans

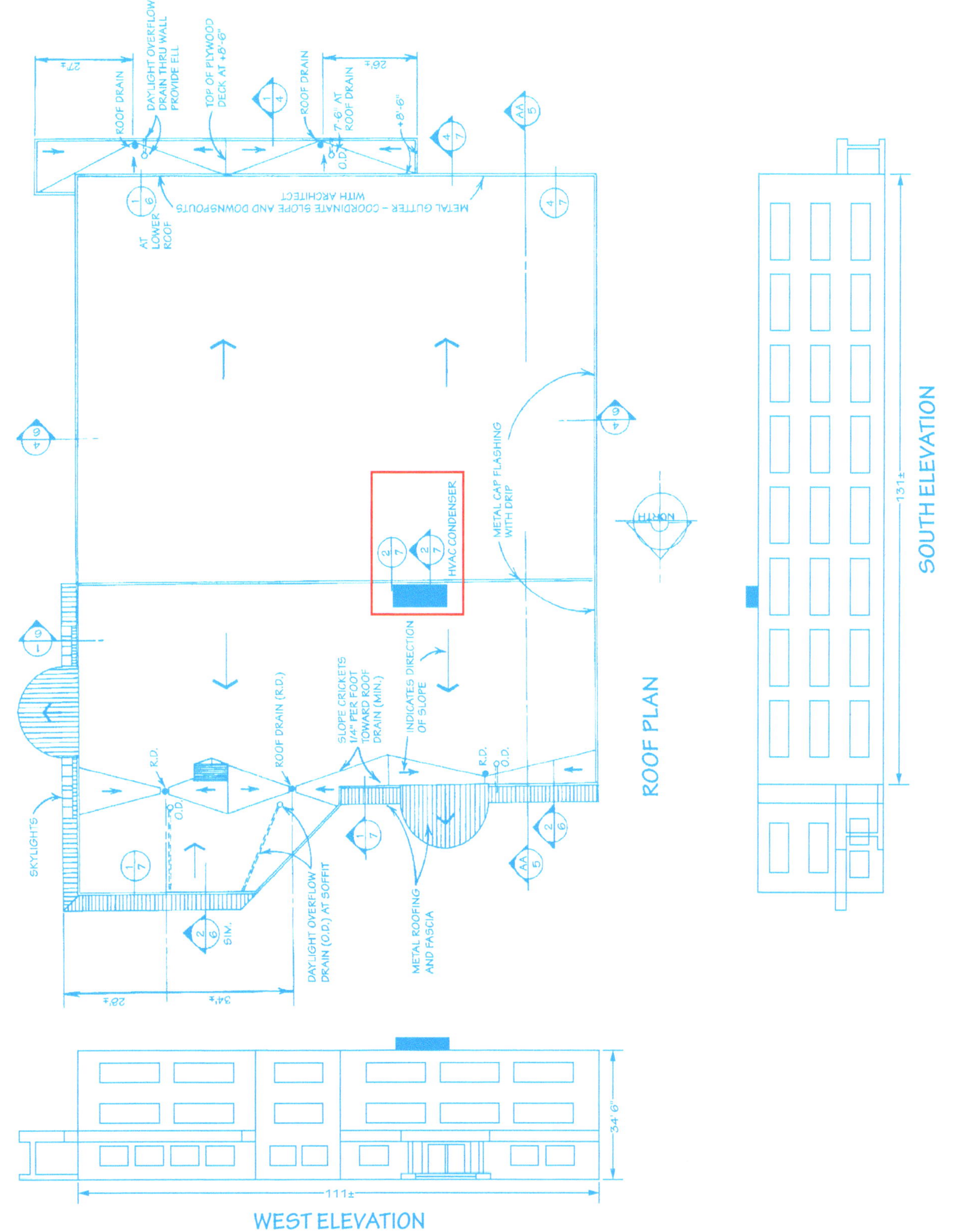

HVAC Condenser Schematics

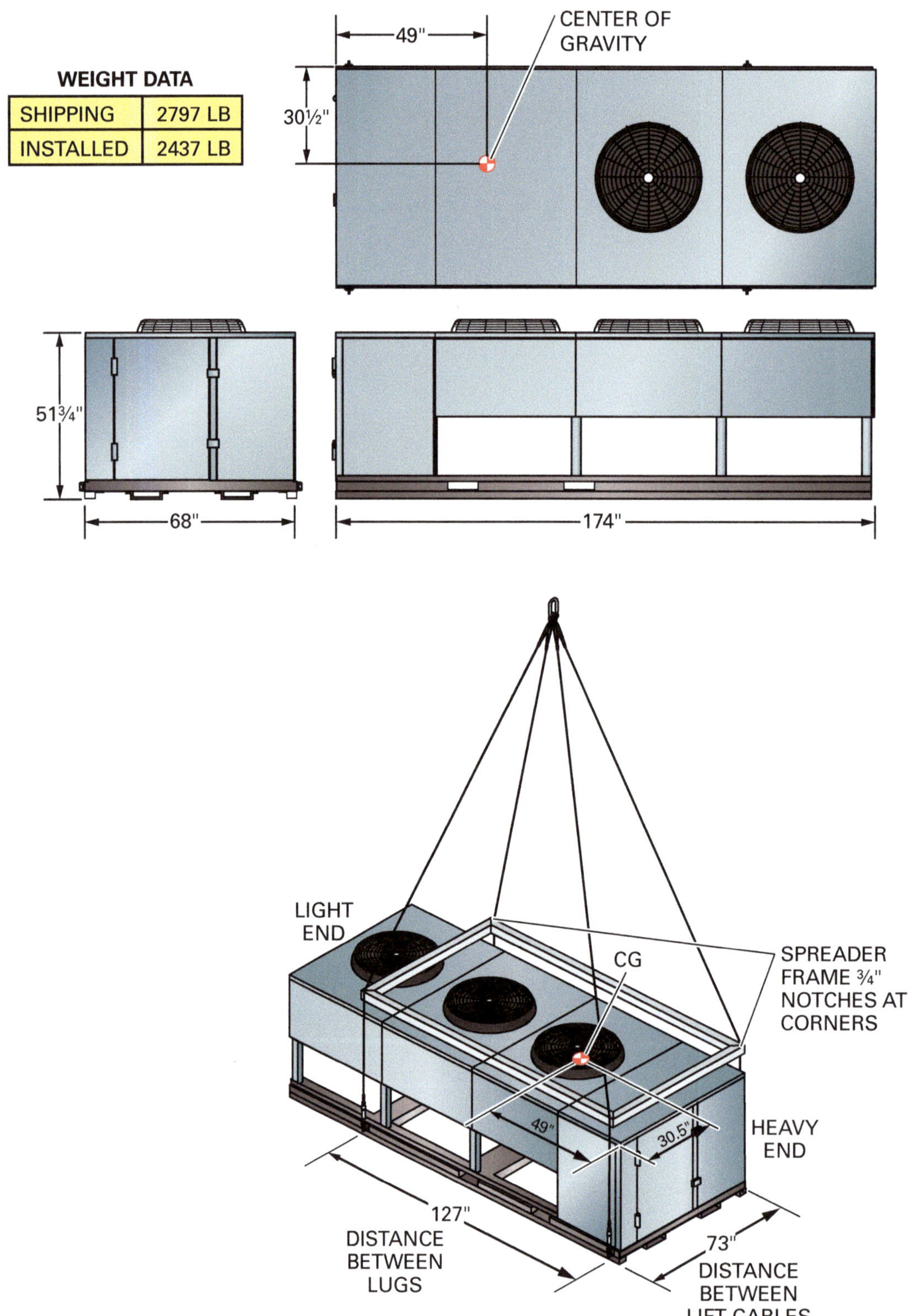

GROVE TM1500 LOAD CHART

WORKING RANGE DIAGRAM
(BOOM DEFLECTION NOT SHOWN)

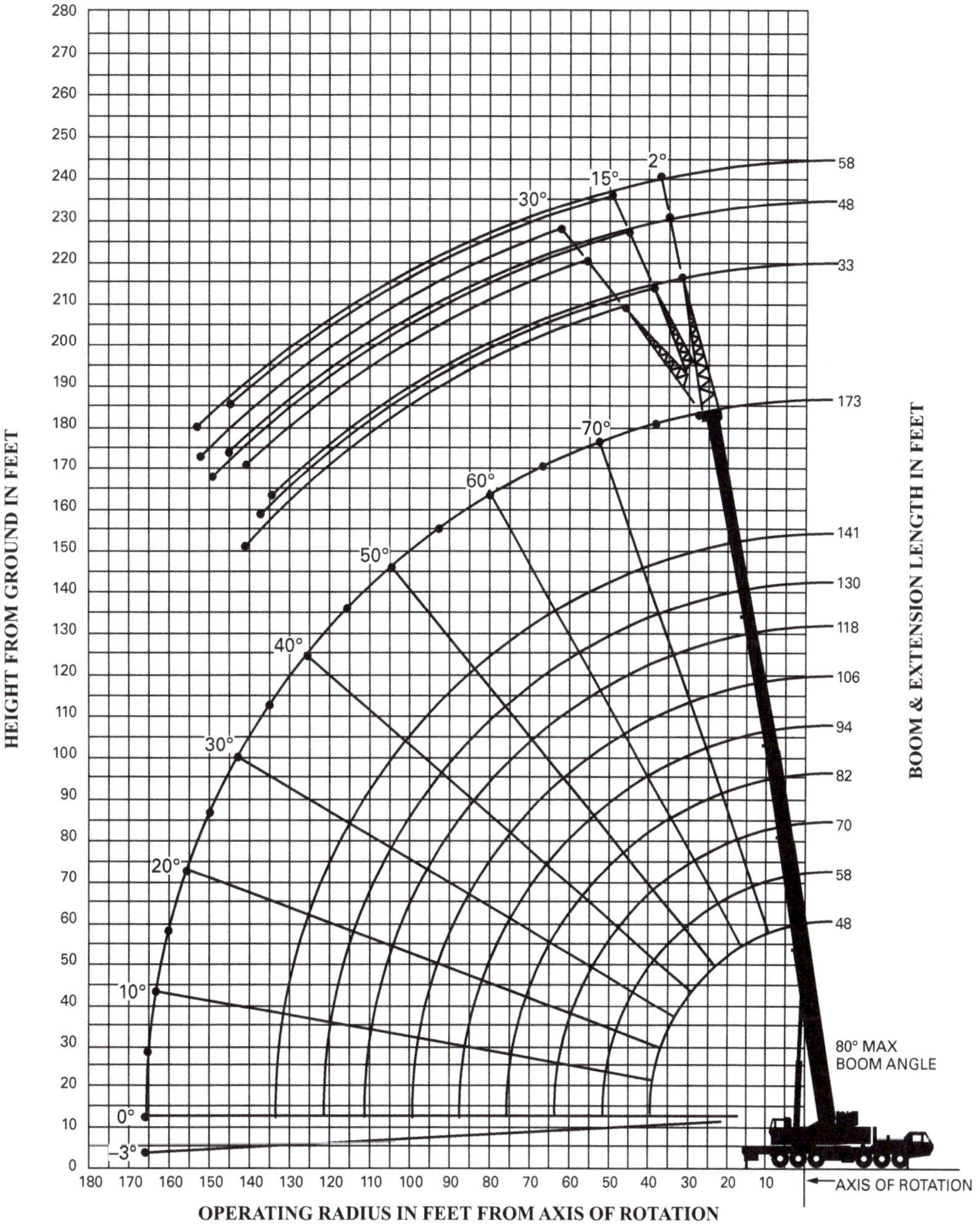

46 FT. - 88 FT. FIXED JIBS ON OUTRIGGERS - 360°

Main Boom Angle (Deg.)	46 FT. JIB						60 FT. JIB					
	#0071		#0072		#0073		#0074		#0075		#0076	
	5° OFFSET		17° OFFSET		30° OFFSET		5° OFFSET		17° OFFSET		30° OFFSET	
	Rad. Ref. (ft.)*	Cap. lbs.**	Rad. Ref. (ft.)*	Cap. lbs.**	Rad. Ref. (ft.)*	Cap. lbs.**	Rad. Ref. (ft.)*	Cap. lbs.**	Rad. Ref. (ft.)*	Cap. lbs.**	Rad. Ref. (ft.)*	Cap. lbs.**
80	44.4	14,900	52.4	12,600	59.8	8,120	47.1	12,500	58.6	8,530	68.7	5,350
77.5	53.7	14,450	61.6	11,650	68.8	7,750	57.1	11,650	68.2	7,850	78.1	5,070
75	62.9	14,000	70.6	10,850	77.6	7,410	67.0	10,800	77.9	7,260	87.3	4,830
72.5	72.0	13,600	79.5	10,150	86.2	7,120	76.8	10,000	87.2	6,760	96.5	4,610
70	81.0	12,470	88.3	9,550	94.8	6,860	86.5	9,230	96.5	6,330	105.3	4,420
67.5	89.8	11,250	96.9	8,980	103.2	6,630	96.0	8,500	105.6	5,950	114.2	4,250
65	98.4	8,850	105.3	7,940	113.2	6,430	105.2	7,850	114.5	5,620	122.7	4,100
62.5	106.9	6,900	113.5	6,550	119.2	6,210	114.3	6,430	123.2	5,270	130.9	3,970
60	115.1	5,370	121.5	5,080	126.9	4,800	123.2	5,010	131.6	4,500	138.9	3,580
55	131.0	3,050	136.8	2,870	141.6	2,700	140.2	2,980	147.7	2,670	154.1	2,370
50	145.9	1,550	151.1	1,460	155.2	1,370	156.2	1,430	162.7	1,260	168.2	1,100

Main Boom Angle (Deg.)	74 FT. JIB						88 FT. JIB					
	#0077		#0078		#0079		#0080		#0081		#0082	
	5° OFFSET		17° OFFSET		30° OFFSET		5° OFFSET		17° OFFSET		30° OFFSET	
	Rad. Ref. (ft.)*	Cap. lbs.**	Rad. Ref. (ft.)*	Cap. lbs.**	Rad. Ref. (ft.)*	Cap. lbs.**	Rad. Ref. (ft.)*	Cap. lbs.**	Rad. Ref. (ft.)*	Cap. lbs.**	Rad. Ref. (ft.)*	Cap. lbs.**
80	50.8	9,470	64.3	6,000	78.0	3,530	52.9	7,300	70.1	3,970	86.0	2,330
77.5	61.2	8,790	74.6	5,470	87.8	3,310	64.1	6,580	80.9	3,590	96.2	2,060
75	71.6	7,980	84.7	5,020	97.3	3,120	75.2	5,950	91.5	3,260	106.3	1,820
72.5	81.9	7,280	94.7	4,630	106.8	2,950	86.1	5,380	102.0	2,980	116.2	1,610
70	92.0	6,610	104.3	4,290	116.0	2,800	96.9	4,800	112.2	2,740	126.0	1,450
67.5	101.9	6,030	114.0	4,000	124.9	2,660	107.5	4,300	122.2	2,530	135.3	1,330
65	111.6	5,500	123.3	3,740	133.7	2,540	117.9	3,870	132.1	2,340	144.6	1,220
62.5	121.1	5,040	132.5	3,510	142.1	2,440	128.0	3,510	141.7	2,180	153.5	1,150
60	130.2	4,050	141.2	3,260	150.2	2,350	138.0	2,870	150.9	1,970	162.1	1,080
55	148.1	2,020	158.2	1,810	165.8	1,600						

A-829-007220

NCCER – *Advanced Rigger*

TM1500

Carrier-mounted hydraulic crane /85% Domestic
46 ft. – 173 ft. aerial power pinned boom

WORKING RANGE DIAGRAM
(BOOM DEFLECTION NOT SHOWN)

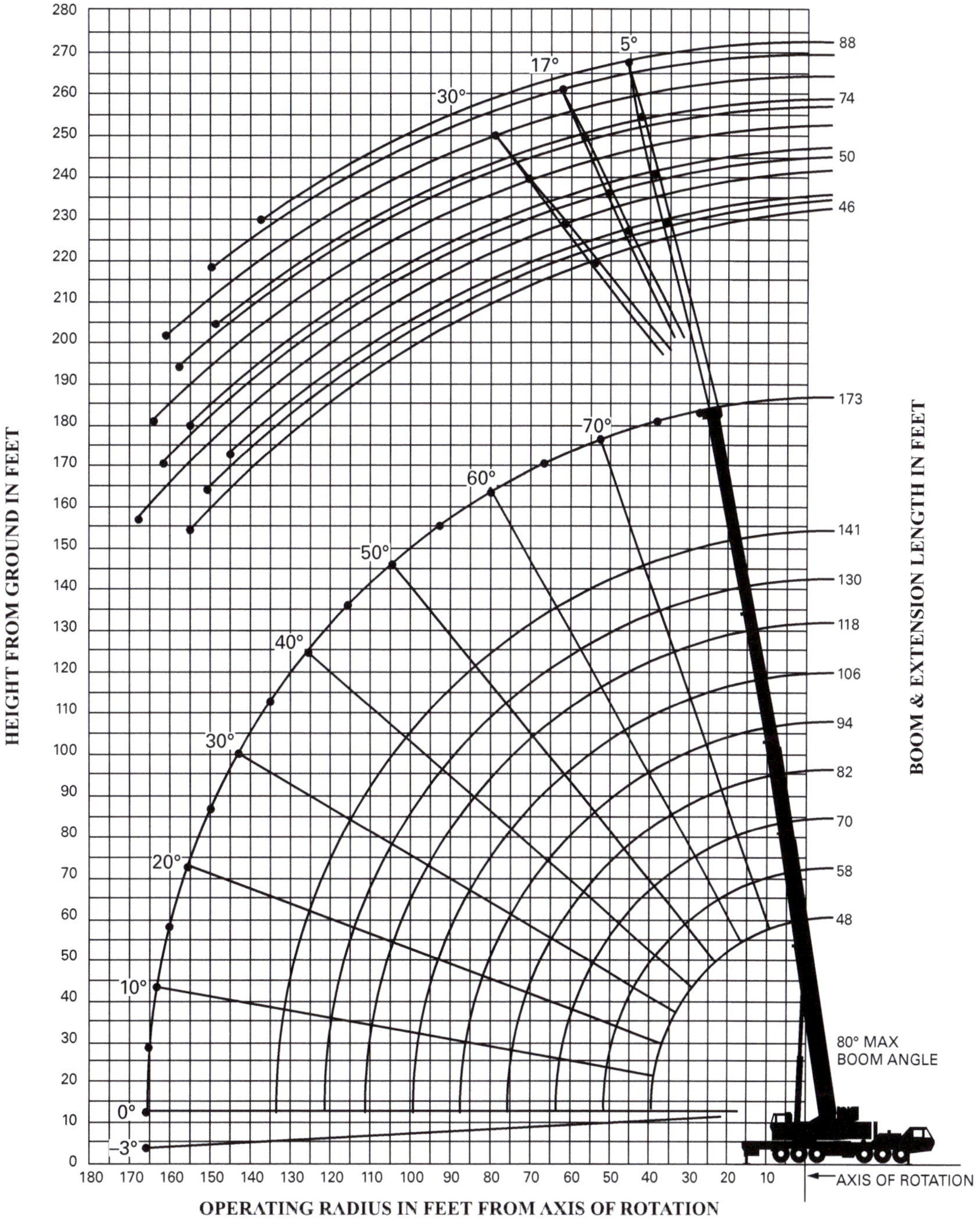

NOTES FOR LIFTING CAPACITIES

WARNING: THIS CHART IS ONLY A GUIDE. The Notes below are for illustration only and should not be relied upon to operate the crane. The individual crane's load chart, operating instructions and other instruction plates must be read and understood prior to operating the crane.

1. All rated loads have been tested to and meet minimum requirements of SAE J1063 0CT80 - Cantilevered Boom Crane Structures - Method of Test, and do not exceed 85% of the tipping load on outriggers (75% of the tipping load on rubber) as determined by SAE J765 OCT 80 Crane Stability Test Code.

2. Capacities given do not include the weight of hookblocks, slings, auxiliary lifting equipment and load handling devices. Their weights MUST be added to the load to be lifted. When more than minimum required reeving is used, the additional rope weight shall be considered part of the load.

3. Capacities appearing above the bold line are based on structural strength and tipping should not be relied upon as a capacity limitation.

4. All capacities are for crane on firm, level surface. It may be necessary to have structural supports under the outrigger floats or tires to spread the load to a larger bearing surface.

5. When either boom length or radius or both are between values listed, the smallest load shown at either the next larger radius or boom length shall be used.

6. Tires shall be inflated to the recommended pressure before lifting on rubber.

7. For outrigger operation, ALL outriggers shall be fully extended with tires raised free of ground before raising the boom or lifting loads.

8. Unless otherwise stated, capacities are with powered boom sections equally extended.

9. With the tele boom extension in working position and main boom length greater than 133 ft. boom angle must not be less than 46.5°, since loss of stability will occur causing a tipping condition.

Constant improvement and engineering progress make it necessary that we reserve the right to make specification, equipment, and price changes without notice. Illustrations shown may include optional equipment and accessories and may not include all standard equipment.

Grove Worldwide - World Headquarters
1565 Buchanan Trail East
Shady Grove, Pennsylvania 17256
Phone: (717) 597-8121 Telex: 1842308 Fax: (717) 597-4062

Grove North America
P.O. Box 21, Shady Grove, Pennsylvania 17256
Western Hemisphere, Asia/Pacific
Phone: (717) 597-8121 Telex: 1842308 Fax: (717) 597-4062

Grove Europe*
Sunderland, England SR4 6TT
Europe, Africa, Middle East, Near East
Phone: (091) 565-6281 Telex: 53484 CRANESG FAX: (091) 564-0442
***Grove Europe Limited, Registered in England, Number 1845128, Registered office, Crown Works, Pallion, Sunderland, Tyne & Wear, England SR4 6TT.**

FORM NO: LC-TMI500-Dom. P/N 3-416 793-3M PRINTED IN U.S.A.

RATED LIFTING CAPACITIES IN POUNDS
46 FT. - 173 FT. BOOM ON OUTRIGGERS - OVER REAR

Radius in Feet	#0001									#0002
	Main Boom Length in Feet (Power Pinned Fly Retracted)									Power Pin Fly Ext. & 141 ft.
	46	58	70	82	94	106	118	130	141	173
10	300,000 (74.5)									
12	280,000 (72)	143,500 (76)	142,000 (79)							
15	235,000 (67.5)	143,500 (72.5)	141,500 (76.5)	130,000 (78.5)						
20	173,500 (60.5)	143,500 (67.5)	123,500 (72)	112,000 (75)	102,000 (77.5)	90,300 (79.5)				
25	136,500 (52)	131,500 (61.5)	110,500 (67.5)	98,650 (71)	89,250 (74)	78,550 (76.5)	73,700 (78.5)	69,300 (80)		
30	106,000 (43)	106,000 (55.5)	98,000 (63)	88,350 (67.5)	78,750 (71)	69,250 (73.5)	65,100 (75)	61,000 (77.5)	60,000 (79.5)	
35	84,700 (30.5)	84,700 (49)	84,700 (58)	80,150 (63.5)	69,000 (67.5)	60,750 (70.5)	57,150 (73)	54,000 (75.5)	52,150 (77.5)	
40		70,500 (41)	70,500 (52.5)	70,500 (59.5)	51,300 (64)	54,000 (67.5)	50,600 (70.5)	48,300 (73)	45,850 (75)	38,000 (79)
45		58,850 (32)	58,850 (47)	58,850 (55)	55,000 (60.5)	48,500 (64.5)	46,200 (68)	43,050 (71)	40,400 (73)	35,750 (77)
50		49,600 (17.5)	49,600 (40.5)	49,600 (50.5)	48,750 (57)	43,050 (61.5)	40,700 (65)	38,250 (68.5)	35,750 (71)	32,100 (75.5)
60			36,200 (22.5)	36,200 (38.5)	36,200 (48.5)	34,300 (55)	33,600 (58.5)	30,750 (63.5)	28,500 (66.5)	26,350 (72)
70				26,050 (25)	26,050 (39.5)	26,050 (47.5)	26,050 (53)	24,750 (58)	23,100 (61.5)	22,000 (68.5)
80					18,850 (27)	18,850 (39)	18,850 (46.5)	18,850 (52.5)	18,700 (56.5)	18,500 (64.5)
90						13,500 (28)	13,500 (38.5)	13,500 (46.5)	13,500 (51.5)	15,250 (60.5)
100							9,390 (29)	9,390 (39)	9,390 (45.5)	12,600 (56.5)
110							6,080 (12.5)	6,080 (30.5)	6,080 (39)	10,100 (52)
120								3,390 (17.5)	3,390 (31)	7,530 (47.5)
130									1,150 (19.5)	5,390 (42.5)
140										3,610 (36.5)
150										2,100 (30)
Minimum boom angle (deg.) for indicated length (no load)									10	19
Maximum boom length (ft) at 0 deg boom angle (no load)									140	167

NOTE: ( ) Boom angles are in degrees.
#LMI operating code. Refer to LMI manual for instructions.

A6-829-007136 & -007143

WEIGHT REDUCTIONS FOR LOAD HANDLING DEVICES

33 ft. Extension	
* Stowed -	892 lbs.
* Erected -	5,704 lbs.

33 ft. - 58 ft. Extension	
* Stowed -	1,246 lbs.
* Erected (Ret.) -	8,412 lbs.
* Erected (Ext.) -	11,406 lbs.

46 ft. - 173 ft. Boom with	
* 46 ft. Jib Erected -	9,613 lbs.
* 60 ft. Jib Erected -	14,571 lbs.
* 74 ft. Jib Erected -	20,443 lbs.
* 88 ft. Jib Erected -	27,199 lbs.
* Fixed Jib Accessories -	327 lbs.

* Reduction of main boom capacities

HOOKBLOCKS:	
30 Ton, 1 Sheave	1,022 lbs.
150 Ton, 8 Sheave	5,254 lbs.
Auxiliary Boom Head	261 lbs.
10 Ton Headache Ball	560 lbs.
15 Ton Headache Ball	803 lbs.

RATED LIFTING CAPACITIES IN POUNDS
46 FT. - 173 FT. BOOM ON OUTRIGGERS - 360°

Radius in Feet	#0001									#0002
	Main Boom Length in Feet (Power Pinned Fly Retracted)									Power Pin Fly Ext. & 141 ft.
	46	58	70	82	94	106	118	130	141	173
10	300,000 (74.5)									
12	280,000 (72)	143,500 (76)	142,000 (79)							
15	235,000 (67.5)	143,500 (72.5)	141,500 (76.5)	130,000 (78.5)						
20	173,500 (60.5)	143,500 (67.5)	123,500 (72)	112,000 (75)	102,000 (77.5)	90,300 (79.5)				
25	135,500 (52)	131,500 (61.5)	110,500 (67.5)	98,650 (71)	89,250 (74)	78,550 (76.5)	73,700 (78.5)	69,300 (80)		
30	106,000 (43)	106,000 (55.5)	98,000 (63)	88,350 (67.5)	78,750 (71)	69,250 (73.5)	65,100 (76)	61,000 (77.5)	60,000 (79.5)	
35	84,700 (30.5)	84,700 (49)	84,700 (58)	80,150 (63.5)	69,000 (67.5)	60,750 (70.5)	57,150 (73)	54,000 (75.5)	52,150 (77.5)	
40		70,500 (41)	70,500 (52.5)	70,500 (59.5)	61,300 (64)	54,000 (67.5)	50,600 (70.5)	48,300 (73)	45,850 (75)	38,000 (79)
45		57,250 (32)	57,250 (47)	57,250 (55)	55,000 (60.5)	48,500 (64.5)	45,200 (68)	43,050 (71)	40,400 (73)	35,750 (77)
50		46,550 (17.5)	46,550 (40.5)	46,550 (50.5)	46,550 (57)	43,050 (61.5)	40,700 (65)	38,250 (68.5)	35,750 (71)	32,100 (75.5)
60			32,000 (22.5)	32,000 (39.5)	32,000 (48.5)	32,000 (55)	32,600 (58.5)	30,750 (63.5)	28,500 (66.5)	26,350 (72)
70				22,550 (25)	22,550 (39.5)	22,550 (47.5)	22,550 (53)	22,550 (58)	22,550 (61.5)	22,000 (68.5)
80					15,900 (27)	15,900 (39)	15,900 (46.5)	15,900 (52.5)	15,900 (56.5)	18,500 (64.5)
90						11,000 (28)	11,000 (38.5)	11,000 (46.5)	11,000 (51.5)	15,250 (60.5)
100							7,280 (29)	7,280 (39)	7,260 (45.5)	11,750 (56.5)
110							4,260 (12.5)	4,260 (30.5)	4,260 (39)	8,520 (52)
120								1,830 (17.5)	1,830 (31)	5,850 (47.5)
130										3,610 (42.5)
140										1,710 (36.5)
Minimum boom angle (deg.) for indicated length (no load)									20	30
Maximum boom length (ft) at 0 dog boom angle (no load)									133	166

NOTE: ( ) Boom angles are in degrees.
#LMI operating code. Refer to LMI manual for instructions.

A6-829-007136 & -007143

33 FT. FIXED LENGTH EXTENSION ON OUTRIGGERS - 360°

Main Boom Angle (Deg.)	#0051		#0052		#0053	
	2° OFFSET		15° OFFSET		30° OFFSET	
	Rad. Ref. (ft.)	Cap. lbs.	Rad. Ref. (ft.)	Cap. lbs.	Rad. Ref. (ft.)	Cap. lbs.
80	39.1	22,850	45.9	20,200	52.0	14,750
75	56.6	17,400	63.0	16,100	68.7	12,750
70	73.7	14,000	79.6	12,950	84.8	10,700
65	90.2	11,850	95.5	10,750	100.3	9,210
60	106.0	10,750	110.8	9,140	115.0	8,050
55	121.1	7,240	125.2	6,160	128.8	5,320
50	135.2	4,330	138.6	3,700	141.8	3,150

A6-829-008157

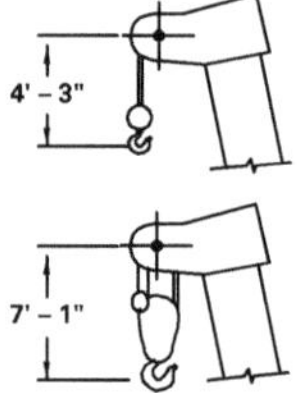

33 FT. - 58 FT. TELE EXTENSION ON OUTRIGGERS - 360°

Main Boom Angle (Deg.)	33 ft. EXTENSION						48 ft. EXTENSION						58 ft. EXTENSION					
	#0021		#0022		#0023		#0031		#0032		#0033		#0041		#0042		#0043	
	2° OFFSET		15° OFFSET		30° OFFSET		2° OFFSET		15° OFFSET		30° OFFSET		2° OFFSET		15° OFFSET		30° OFFSET	
	Rad. Ref. (ft.)	Cap. lbs.	Rad. Ref. (ft.)	Cap. lbs.	Rad. Ref. (ft.)	Cap. lbs.	Rad. Ref. (ft.)	Cap. lbs.	Rad. Ref. (ft.)	Cap. lbs.	Rad. Ref. (ft.)	Cap. lbs.	Rad. Ref. (ft.)	Cap. lbs.	Rad. Ref. (ft.)	Cap. lbs.	Rad. Ref. (ft.)	Cap. lbs.
80	39.1	22,300	45.9	19,650	52.0	14,200	43.9	14,800	55.8	12,700	65.0	9,040	45.1	9,600	59.1	8,300	69.6	7,000
75	56.6	16,850	63.0	15,550	68.7	12,200	62.5	13,700	73.4	10,900	81.8	8,320	65.0	9,150	77.9	7,950	87.8	6,440
70	73.7	13,450	78.6	12,400	84.8	10,150	80.7	11,600	90.3	8,840	97.9	7,110	84.4	8,550	96.2	7,200	105.5	5,760
65	90.2	11,300	95.5	10,200	100.3	8,870	98.2	9,230	106.7	7,330	113.3	6,120	103.2	7,630	113.8	6,000	122.3	4,970
60	106.0	10,200	110.8	8,600	115.0	7,450	115.1	7,550	122.1	8,210	127.9	5,340	121.2	6,260	130.5	5,100	138.2	4,380
55	121.1	6,400	125.2	5,370	128.8	4,540	131.1	5,200	136.7	3,860	141.5	2,950	138.2	4,650	146.2	3,265	153.1	2,380
50	135.2	3,520	138.6	2,810	141.8	2,360	146.1	2,580	150.2	1,790			154.3	2,235				

#LMI operating code. Refer to LMI manual for instructions.

A6-829-007158

NCCER – *Advanced Rigger*

LIFT DATA SHEET

Lift Data Sheet (Single Mobile Crane Lift)   Date:

Load Name: ______   Lift Description: ______

Project: ______   Units: U.S. (ft, lb)

Crane Details   Manufacturer: ______   Model Number: ______

Configuration: ______	Base Mount Type: ______	Track/Outrigger Ext: ______ ft
Boom Type: ______	Boom Length Used: ______ ft	Boom/Jib Angle: ______ deg.
Jib Type: ______	Jib Length Used: ______ ft	Tail Swing: ______ ft
Crane Ballast: ______ lb	Aux Counterweight ______ lb	
Block Capacity: ______ ton	Line Size ______ in	
Single Line Pull: ______ lb	Parts Line Used ______	Reeved Capacity: ______ lb
Heavy Lift Attachments: ______	Load Radius: ______ ft	Superlift Weight: ______ ton

Load Details

	Quantity	Wt. ea.	Total Weight
Basic Weight of Load/Item	______	______ lb	______ lb
______		______ lb	______ lb
______		______ lb	______ lb
______		______ lb	______ lb
______		______ lb	______ lb
		Total Weight of Load/Item:	______ lb

Rigging Data (size, type, and capacity)

	Wt. ea.	Total Weight
______	______ lb	______ lb
______	______ lb	______ lb
______	______ lb	______ lb
______	______ lb	______ lb
______	______ lb	______ lb
______	______ lb	______ lb
______	______ lb	______ lb
______	______ lb	______ lb
	Total Rigging Weight:	______ lb

Additional Weight Items

	Wt. ea.	Total Weight
Main Hook	______ lb	______ lb
Wire Rope	______ lb	______ lb
Other Suspended Hooks	______ lb	______ lb
Aux Boom Sheaves	______ lb	______ lb
Jib	______ lb	______ lb
______	______ lb	______ lb
______	______ lb	______ lb
	Total Additional Weight:	______ lb
	GROSS LOAD (sum of all items above):	8,991 lb

Crane Capacities   Condition: Placing Condenser

Total Load to Crane:	8,991 lb	______ lb	______ lb
Planned Radius:	73 ft	______ ft	______ ft
Chart Radius Used:	80 ft	______ ft	______ ft
Chart Capacity:	18,850 lb	______ lb	______ lb
% of Chart Capacity:	47.7 %	______ %	______ %

MAXIMUM PERCENTAGE OF CHART CAPACITY TO BE USED IS   47.7 %

Total Suspended Load: 3,476 lb   Reeved Capacity: 51,800 lb

MAXIMUM PERCENTAGE OF REEVED CAPACITY PLANNED TO BE USED IS   6.7 %

Document Attachments

○ Crane Layout	○ Lift Plan Category	○ GBP Source/Calculations
○ Rigging Hookup Diagram	○ AHA	○ Project Scope
○ Crane Load Chart	○ Risk Evaluation	○ Load Weight/CG Source
○ Foundation Details	○ Weather Forecast	○ Load/Crane Clearances
○ Crane Cribbing Arrangement	○ Load Schematic	○

Notes ______

	Name (Print)	Signature	Title	Date
Prepared by:	______	______	______	______
Checked by:	______	______	______	______
Approved:	______	______	______	______

CRITICAL LIFT PLAN

CRITICAL LIFT PLAN

DESCRIPTION OF WORK · SECTION 1A

Date:	Time:
Associated Work Permit Numbers:	
JOB Number:	
Location Name:	
Street Address/Building Number:	
City / State:	
Weather:	Wind Speed: · Visibility:
Load Description:	
Task:	
Estimated Duration for Task:	
Crane Make & Model:	Crane Operator:

Lift Supervisor Name [print]: _________________________________   Phone: (_______)_______-________

HAZARDS IDENTIFIED

- ☐ OVERHEAD UTILITIES / OBSTRUCTION
- ☐ GROUND CONDITIONS
- ☐ UNDERGROUND UTILITIES
- ☐ VEHICLE & PEDESTRIAN TRAFFIC
- ☐ PINCH POINTS
- ☐ OTHER EQUIPMENT
- ☐ FALL HAZARDS
- ☐ OTHER: _______________________________
  ____

PRECAUTIONS

- ☐ SPOTTER
- ☐ ELEVATED WARNING SIGNS
- ☐ CRIBBING / SHORING / TRENCH PLATES
- ☐ K-RAILS / JERSEY BARRIER
- ☐ CAUTION TAPE / BARRICADE
- ☐ FALL PROTECTION
- ☐ OTHER: _______________________________
  ____
  ____
  ____

Version 2c: 10/16/12

SAFETY

☐ Slings & Rigging Inspected **PRIOR TO USE**

☐ Objects in eyelets of synthetic slings **DO NOT** exceed 1/3 the length of the eyelet

☐ Objects in eyelets of wire rope slings **DO NOT** exceed 1/2 the length of the eyelet

☐ The D/d Ratio of any attached rigging equipment to the Wire Rope is 25:1

☐ The Included Angle of the slings attaching to the hook **DOES NOT** exceed 90°

☐ Sharp Edge Protection

☐ Shield and protect slings and rigging against chemical exposure and extreme heat

☐ Safety notified 5 days prior to all lifts

SLINGS

☐ SYNTHETIC ☐ WIRE ROPE
☐ CHAIN ☐ WIRE MESH

SLING CONFIGURATION & CAPACITY

☐ Vertical: ________________________
[Vertical is within 0° to 5° to be 100%]

☐ Basket: ________________________
[Wire Rope Slings require D/d of 25:1]

☐ Choker: ________________________
[Starting @ approximately 75% of Vertical]

Number of Slings: ________________________

☐ 2 Leg Bridle
☐ 3 Leg Bridle
☐ 4 Leg Bridle
☐ More than 4 Leg Bridle
☐ Tag Line

SHACKLES

☐ Shackles are at 100% of capacity when slings are from 0° to 5° of vertical
☐ Shackles are at 70% of capacity when the included sling angles are from 12° to 90° of vertical
☐ Shackles are at 50% of capacity when the included sling angles are from 91° to 180°

Number of Shackles: ________________________

Shackle Capacity: ________________________

EYEBOLTS

☐ Non-shouldered eyebolts are at 100% of capacity when the slings are from 0° to 5° of vertical
☐ Non-shouldered eyebolts are **NOT** recommended with sling angles more than 5° from vertical
☐ Shouldered eyebolts are at 100% of capacity when the slings are from 0° to 5° of vertical
☐ Shouldered eyebolts are at 70% of capacity when the sling angles are from 6° to 45° of vertical
☐ Shouldered eyebolts are at 50% of capacity when he sling angles are from 46° to 90° of vertical

☐ Shouldered   ☐ Non-shouldered

Eyebolt Capacity: ________________________ (lbs.)

HOOK

Included angle to the hook **DOES NOT** exceed 90°

☐ Latch required
☐ Latch moused in the open position
☐ Latch moused in the closed position

MASTERLINK

Masterlink Capacity: ________________________

SWIVEL HOIST RING

Torque value: ________________________

Swivel hoist ring capacity:

LIST ALL ADDITIONAL EQUIPMENT AND INFORMATION NOT COVERED IN SECTION 1B

☐ ____

☐ ____

☐ ____

☐ ____

☐ ____

☐ ____

☐ ____

☐ ____

☐ ____

☐ ____

☐ ____

Version 2c: 10/16/12

LOAD #1

Width: __________   Length: __________

Height: _________   Weight: __________

Number of lift points: _________

Number of slings: ______   Length of Sling Leg: ______

Distance Between Opposing Sling: ________________

Height Between Hook and Load Plane: ___________

Sling Tension:

LOAD #2

Width: __________   Length: __________

Height: _________   Weight: __________

Number of lift points: _________

Number of slings: ______   Length of Sling Leg: ______

Distance Between Opposing Sling: ________________

Height Between Hook and Load Plane: ___________

Sling Tension:

LOAD #3

Width: __________   Length: __________

Height: _________   Weight: __________

Number of lift points: _________

Number of slings: ______   Length of Sling Leg: ______

Distance Between Opposing Sling: ________________

Height Between Hook and Load Plane: ___________

Sling Tension:

LOAD #4

Width: __________   Length: __________

Height: _________   Weight: __________

Number of lift points: _________

Number of slings: ______   Length of Sling Leg: ______

Distance Between Opposing Sling: ________________

Height Between Hook and Load Plane: ___________

Sling Tension:

LOAD #5

Width: __________   Length: __________

Height: _________   Weight: __________

Number of lift points: _________

Number of slings: ______   Length of Sling Leg: ______

Distance Between Opposing Sling: ________________

Height Between Hook and Load Plane: ___________

Sling Tension:

SLING TENSION FORMULA

LOAD ON SLING CALCULATED
TENSION 1 = LOAD X D2 X S1/(H(D1+D2))
TENSION 2 = LOAD X D1 X S2/(H(D1+D2))

USE ADDITIONAL PAGES AS NEEDED

Version 2c: 10/16/12

DIAGRAM OF LOAD [INCLUDE MEASUREMENTS]

DIAGRAM OF PROPOSED RIGGING & SLINGS [INCLUDE MEASUREMENTS]

Version 2c: 10/16/12

<table><tr><td>**MAN BASKET / PERSONNEL PLATFORM**</td><td>**SECTION 1F**</td></tr></table>

OSHA 29 CFR 1926.1400 ; NASA-STD-8719.9

SAFETY

Types of Communication: ☐ Hand ☐ Voice ☐ Radio ☐ Hardline

☐ **CRANE / WIRE ROPE CAPACITY DE-RATED 50%**   ☐ **WINDS UNDER 20MPH**

☐ Over Water
  Depth of Water: _____________________   Height to Water: _______________
☐ Personal Floatation Device Required   ☐ Fall Harness Required

☐ Basket & Rigging Inspected   ☐ 125% Capacity Test Pick at each radius

CRANE INFORMATION

Crane Operator Name: _____________________   Crane Make & Model: _________________________

Outrigger / Crawler Configuration:
  ☐ On Rubber   ☐ 0% Extended   ☐ 50% Extended   ☐ 100% Extended

CAPACITY INFORMATION

☐ **CRANE CAPACITY DE-RATED 50%**   ☐ **WIRE ROPE CAPACITY DE-RATED 50%**

PICK #1

Main Boom Length & Configuration: ______________   Jib Length & Configuration: __________________

Radius: ______ Crane Cap: __________ Wire Rope Cap: __________ **Capacity LESS 50%:** __________

PICK #2

Main Boom Length & Configuration: ______________   Jib Length & Configuration: __________________

Radius: ______ Crane Cap: __________ Wire Rope Cap: __________ **Capacity LESS 50%:** __________

NAMES & WEIGHTS OF PERSONNEL IN BASKET

Name: ____________________WT: ______   Name: ____________________WT: ______

Name: ____________________WT: ______   Name: ____________________WT: ______

Rated Capacity of the Man Basket: _________   Total Weight of Men in Basket: _____________
                                               Weight of Rigging: _____________
☐ 125% Capacity Test Pick at each radius   _________________________________
                                               TOTAL WEIGHT:

OSHA 29 CFR 1926.1400 ; NASA-STD-8719.9

Utility Owner: ________________________ Utility Contact: ________________________

Contact Phone Number: ________________________

Verified Voltage of Power Lines: ________________ Verified Elevation of Power Lines: ______

Required **MINIMUM** Clearance (see table below): ________________________

Cranes Working Radius: ________________________

☐ DE-ENERGIZED & GROUNDED

☐ LOCKED OUT / TAGGED OUT
☐ Verified By: ____
                    [printed name]

☐ ENERGIZED

☐ Insulating Link / Device
☐ Non-Conductive Slings
☐ Non-Conductive Tag Line
☐ Distance from Crane to Power Line: ________________
☐ Distance from Crane Boom Tip to Power Line: __________

WARNING DEVICES

☐ Proximity Alarm
☐ Elevated Warning Signs
☐ Spotter

REQUIRED MINIMUM CLEARANCES

```
0—50kV....................10'
50—200kV.................15"
200—350kV...............20'
350—500kV...............25'
500—750kV...............35'
750—1,000kV.............45'
1,000kV +.................Determined by Registered Engineer / Utility Owner
```

Version 2c: 10/16/12

SAFETY

- ☐ **SAFETY LIFT OBSERVER PRESENT**   NAME OF OBSERVER: ________________________
- ☐ **HOIST LINES VERTICAL**
- ☐ **NO SIDE LOADING**
- ☐ **CRANES DE-RATED 25%**
- ☐ **ALL CRANES ARE LEVEL**

CRANE # 1 INFORMATION

Crane Operator Name: ____________________   NCCCO License #: ____________________

Crane Make & Model: ____

Outrigger / Crawler Configurations:
 ☐ On Rubber   ☐ 0% Extended   ☐ 50% Extended   ☐ 100% Extended

Main Boom Length & Configuration: _________   Jib Length & Configuration: ____________________

Furthest Operating Radius: ______________   Rated Capacity at Furthest Radius **LESS 25%**: _______

CRANE #2 INFORMATION

Crane Operator Name: ____________________   NCCCO License #: ____________________

Crane Make & Model: ____

Outrigger / Crawler Configurations:
☐ On Rubber   ☐ 0% Extended   ☐ 50% Extended   ☐ 100% Extended

Main Boom Length & Configuration: _________   Jib Length & Configuration: ____________________

Furthest Operating Radius: ______________   Rated Capacity at Furthest Radius **LESS 25%**: _______

CRANE #3 INFORMATION

Crane Operator Name: ____________________   NCCCO License #: ____________________

Crane Make & Model: ____

Outrigger / Crawler Configurations:
☐ On Rubber   ☐ 0% Extended   ☐ 50% Extended   ☐ 100% Extended

Main Boom Length & Configuration: _________   Jib Length & Configuration: ____________________

Furthest Operating Radius: ______________   Rated Capacity at Furthest Radius **LESS 25%**: _______

Version 2c: 10/16/12

Crane Operator Name: _______________________   Crane Make & Model: _______________________

Outrigger / Crawler Configuration:
☐ On Rubber   ☐ 0% Extended   ☐ 50% Extended   ☐ 100% Extended

---

PICK #1

Main Boom Length & Configuration: _____________   Jib Length & Configuration: _________________

Radius: _____   Crane Cap: _________   Wire Rope Cap: _________   Load Wt: _________

PICK #2

Main Boom Length & Configuration: _____________   Jib Length & Configuration: _________________

Radius: _____   Crane Cap: _________   Wire Rope Cap: _________   Load Wt: _________

PICK #3

Main Boom Length & Configuration: _____________   Jib Length & Configuration: _________________

Radius: _____   Crane Cap: _________   Wire Rope Cap: _________   Load Wt: _________

PICK #4

Main Boom Length & Configuration: _____________   Jib Length & Configuration: _________________

Radius: _____   Crane Cap: _________   Wire Rope Cap: _________   Load Wt: _________

PICK #5

Main Boom Length & Configuration: _____________   Jib Length & Configuration: _________________

Radius: _____   Crane Cap: _________   Wire Rope Cap: _________   Load Wt: _________

PICK #6

Main Boom Length & Configuration: _____________   Jib Length & Configuration: _________________

Radius: _____   Crane Cap: _________   Wire Rope Cap: _________   Load Wt: _________

Version 2c: 10/16/12

CRITICAL LIFT PLAN

Lifting Device and Equipment Manger

Safety Concerns/Comments:

Version 2c: 10/16/12

CRITICAL LIFT PLAN

Notifications: If lift is to occur in or adjacent to a facility, notify the appropriate FSM (or functional equivalent) at the lift site and nearby facilities.

LDEM or Alternate

Name (Printed)	Name (Signature)	Date

Project Lift Manager Approval

Name (Printed)	Name (Signature)	Date

Safety Lift Observer (SLO)

Name (Printed)	Name (Signature)	Date

FSM (if lift is in facility)

Name (Printed)	Name (Signature)	Date

Engineer

Name (Printed)	Name (Signature)	Date

Operator

Name (Printed)	Name (Signature)	Date

Rigger

Name (Printed)	Name (Signature)	Date

Name (Printed)	Name (Signature)	Date

Name (Printed)	Name (Signature)	Date

Version 2c: 10/16/12

CRITICAL LIFT PLAN

1. ☐ Verify crane and equipment are within inspection and testing intervals for operation. Review crane history for recent discrepancies, as well as required repairs or modifications that could affect the planned lift. Such history review should typically assess the 12 months prior to the lift.

2. ☐ Verify that rigging equipment to be used is within inspection and test intervals for operation (Include all load-bearing elements such as shackles, turnbuckles, and eyebolts.) and that it has a thirteen digit tag.

3. ☐ Verify operator(s) has current certifications for operation and has received specific operational training instructions on the devices and equipment to be used for the lift.

4. ☐ Identify approved variance (deviation or waiver) if applicable.

5. ☐ Establish safety zones and install appropriate barriers.

6. ☐ Designate cognizant individual to be positioned at the crane's main electrical disconnect (for facility cranes & hoists), to provide alternate emergency stop (required for Critical Lifts).

7. ☐ Assemble operations crew and conduct pre-lift briefing. The briefing should emphasize such details as lift signals, emergency procedures, as well as that emergency stop signal intended to immediately stop all crane operations and crane motions. It should be reiterated that only personnel with identified tasks are allowed in the lift zone, and then only with required personal protective equipment (PPE), proper training, etc.

8. The Operator is to conduct the following "daily inspection": (NOTE: Document all deficiencies. Correct all hazardous deficiencies before using the equipment.)
   - a. ☐ Check functional operation of all control mechanisms for proper action and/or response.
   - b. ☐ Without disassembling anything, visually inspect lines, tanks, valves, drain pumps, gear casings, and other components of fluid systems for deterioration and leaks. This applies to components that you can see from ground level or safely check from crane inspection walkways.
   - c. ☐ Without disassembling anything, visually inspect functional operating and control mechanisms for excessive wear and contamination by excessive lubricants or foreign matter.
   - d. ☐ Visually inspect the hook, latch and swivel for proper operation, cracks, and deformities.
   - e. ☐ Visually (without climbing up to bridge) inspect rope reeving for proper travel and drum lay, and inspect wire rope for obvious kinks, deformations, or damage.
   - f. ☐ Visually inspect hoist chain (where applicable) for excessive wear or distortion.
   - g. ☐ Verify operation of the upper limit switch with no load on the hook.
   - h. ☐ Verify that any multiple lines (ropes) are not twisted around each other, and that the hook is centered over the load before lifting.
   - i. ☐ Verify communications.
   - j. ☐ Verify that all clearances, especially from electrical equipment and power lines, as well as weather conditions, etc., are acceptable.
   - k. ☐ Ensure the load is properly rigged and securely attached to the hook. Use tag lines when required to minimize and control unwanted motion or rotation of the load.
   - l. ☐ Slowly lift the load just an inch or two and hold in suspension for a few moments. Allow the load dynamics to dampen and make sure the load is in balance. Verify that the brake is holding.

9. If off-site travel is required, verify that necessary actions and preparations have been made to travel the roadway, such as obtaining permission, permits, arranging for escorts, etc.

WARNING: KNOW AND OBEY ALL EMERGENCY STOP SIGNALS

Version 2c: 10/16/12

Additional Resources

This module is intended as a thorough resource for task training. The following reference materials are recommended for further study.

Planning for Load Handling Activities,ASME P30.1. New York: The American Society of Mechanical Engineers.

"Load Charts for Use With Written Examinations—Grove TM1500." Manitowoc Crane Group and NCCER. **www.nccer.org/docs/default-source/Load-Charts/grovetm1500.pdf?sfvrsn=853e0e4f_0**

29 *CFR* 1926, Subpart CC,*Cranes and Derricks in Construction.* **www.ecfr.gov**

Figure Credits

Courtesy of DOE/NREL, Credit - Pat Corkery, Module Opener

© iStock.com/BanksPhotos, Figure 1

© iStock.com/Knaupe, Figure 2

SCE&G® A SCANA Company, Figure 3

© iStock.com/Paul Vasarhelyi, Figure 4

© iStock.com/pictafolio, Figure 5

The Manitowoc Company, Inc., Figures 8, 9, Appendix C

© iStock.com/MichaelKnudsen, Figure 13

© Gary Whitton/Shutterstock.com, Figure 15

National Oceanic and Atmospheric Administration/Department of Commerce/NOAA Weather in Focus Photo Contest 2015/Photographer: Jeff Stillman, Figure 17

NASA, Appendix E

Section Review Answer Key

Answer	Section Reference	Objective
Section One		
1. b	1.1.1	1a
2. d	1.2.2	1b
3. c	1.3.1	1c

NCCER CURRICULA — USER UPDATE

NCCER makes every effort to keep its textbooks up-to-date and free of technical errors. We appreciate your help in this process. If you find an error, a typographical mistake, or an inaccuracy in NCCER's curricula, please fill out this form (or a photocopy), or complete the online form at **www.nccer.org/olf**. Be sure to include the exact module ID number, page number, a detailed description, and your recommended correction. Your input will be brought to the attention of the Authoring Team. Thank you for your assistance.

Instructors – If you have an idea for improving this textbook, or have found that additional materials were necessary to teach this module effectively, please let us know so that we may present your suggestions to the Authoring Team.

NCCER Product Development and Revision

13614 Progress Blvd., Alachua, FL 32615

Email: curriculum@nccer.org
Online: www.nccer.org/olf

❏ Trainee Guide   ❏ Lesson Plans   ❏ Exam   ❏ PowerPoints   Other _______________________

Craft / Level: ____   Copyright Date: __________

Module ID Number / Title: ____

Section Number(s): ____

Description: ____

____

____

____

Recommended Correction: ____

____

____

____

Your Name: ____

Address: ____

Email: ____   Phone: ______________________

Hoisting Personnel

OVERVIEW

Using a crane to hoist personnel is potentially hazardous. Loss of control of the crane or the failure of a hoisting component will likely result in the serious injury or death of the workers being hoisted. It was once common practice for workers to ride a ball as a convenient means of moving between the ground and a high point. However, the potential for catastrophic injuries was extreme, and falls and injuries were common. For this reason, the government enacted strict safety regulations pertaining to lifting personnel by crane. This module covers the details of these regulations.

Module 21305

Trainees with successful module completions may be eligible for credentialing through the NCCER Registry. To learn more, go to **www.nccer.org** or contact us at 1.888.622.3720. Our website has information on the latest product releases and training, as well as online versions of our *Cornerstone* magazine and Pearson's product catalog.

Your feedback is welcome. You may email your comments to **curriculum@nccer.org**, send general comments and inquiries to **info@nccer.org**, or fill in the User Update form at the back of this module.

This information is general in nature and intended for training purposes only. Actual performance of activities described in this manual requires compliance with all applicable operating, service, maintenance, and safety procedures under the direction of qualified personnel. References in this manual to patented or proprietary devices do not constitute a recommendation of their use.

Objective

When you have completed this module, you will be able to do the following:

1. Identify and describe the requirements and operational techniques related to personnel hoisting.
   a. Describe the required characteristics of personnel platforms.
   b. Identify the testing and inspection requirements for personnel platforms.
   c. Identify and discuss the operational requirements involved in personnel lifts.

Performance Task

Under the supervision of your instructor, you should be able to do the following:

1. Complete a personnel lift plan using the form in *Appendix C* and a provided scenario.

Trade Terms

Bridle slings
Flemish eyes
Nondestructive testing (NDT)

Poured socket
Proof test
Trial lift

Industry Recognized Credentials

If you are training through an NCCER-accredited sponsor, you may be eligible for credentials from NCCER's Registry. The ID number for this module is 21305. Note that this module may have been used in other NCCER curricula and may apply to other level completions. Contact NCCER's Registry at 888.622.3720 or go to **www.nccer.org** for more information.

Contents

Figures and Tables

1.0.0 HOISTING PERSONNEL

Objective

Identify and describe the requirements and operational techniques related to personnel hoisting.

a. Describe the required characteristics of personnel platforms.
b. Identify the testing and inspection requirements for personnel platforms.
c. Identify and discuss the operational requirements involved in personnel lifts.

Performance Task

Complete a personnel lift plan using the form in *Appendix C* and a provided scenario.

Trade Terms

Bridle slings: Lift rigging consisting of multi-leg, wire rope, or chain slings.

Flemish eyes: Spliced loops formed in the end of wire rope rigging slings. For personnel hoist rigging, the eyes must also contain a thimble and be mechanically crimped with a ferrule.

Nondestructive testing (NDT): Testing the integrity of fabricated metal structures, especially welds, by various means to detect hidden flaws without destroying the object or rendering it useless. Common NDT methods include dye penetrant (PT), magnetic particle (MT), X-ray and radiography (RT), and ultrasound (UT).

Poured socket: Referring to wire rope terminations, a method of attaching a fitting by placing the end of the rope inside the fitting socket, then pouring molten zinc, another metal alloy, or a special resin into the socket. After hardening, these terminations can provide 100 percent of the wire rope's minimum breaking strength. This process is also referred to as speltering.

Proof test: A strength test of an object, usually under a load exceeding that expected under normal use, to ensure it is suitable for safe use. In personnel hoisting, a proof test requires loading the platform with a weight equal to 125 percent of its rated load and suspending it for at least 5 minutes.

Trial lift: For a personnel hoisting platform, a lift of the platform and rigging without occupants, but with an equivalent load, under the same conditions and over the same route as the planned personnel lift.

Two key standards address personnel hoisting platforms, often called *manbaskets*. For the design, manufacturing, and operation of such equipment, *ASME B30.23, Personnel Lifting Systems*, provides technical details. 29 *CFR* 1926.1431, *Hoisting Personnel*, provides the safety regulations for hoisting personnel with cranes. The OSHA standard incorporates the relevant parts of the ASME consensus standard and makes them enforceable laws.

29 *CFR* 1926.1431 prohibits the hoisting of personnel except when the employer demonstrates that workers can't complete their tasks by any other means. The OSHA standard notes that Subpart R (rather than 1926.1431) covers hoisting requirements specific to steel erection.

Some form of hoisting platform meeting the requirements of the standard is necessary for all applicable lifts *except*:

- Into or out of drill shafts less than 8 feet (2.44 m) in diameter
- Pile-driving operations
- Personnel transfers at marine worksites (*Figure 1*)
- In storage tanks, shafts, and chimneys

The OSHA standard provides specific requirements applicable to these situations.

1.1.0 Platform Characteristics

ASME B30.23 provides detailed requirements for the design, construction, testing, and inspection of personnel lifting platforms. It also includes operational guidelines to ensure the safe operation of such platforms. 29 *CFR* 1926.1431 incorporates the essential ASME standard's platform characteristics relating to occupant safety.

1.1.1 Design

The first design concern addressed by the ASME standard is the design load factor. For a platform equipped for a single-point lift, the attachment point must have a design factor of 7. This means that it must be able to support seven times the

Figure 1 A personnel lift at a marine work site.

design load of the platform's occupants and materials, including the platform's weight. For platforms with two or more hoisting attachments, any loaded pair of attachment points must have a design factor of 5.

Besides the requirement for a qualified person familiar with structural design to design platforms, design features for a personnel lift platform must include the following:

- Minimum capacity of 300 pounds (136 kg)
- A guardrail consisting of a top rail, intermediate rail, lower barrier, and toe board of specified dimensions. The barrier may be solid, perforated, or of expanded metal mesh with openings no larger than $\frac{1}{2}$ inch (13 mm).

- All structural components of the platform enclosure capable of resisting specified minimum downward and outward forces to ensure security of the platform occupants.
- Each anchor point for attaching fall protection devices capable of stopping the combined weight of all allowed occupants falling through 6 feet (1.8 m).
- Hand railing with a specified standoff along entire perimeter of enclosure except at access doors/gates.
- Slip-resistant decking with drain holes.
- Means to secure loose tools, materials, and other items within the enclosure.
- A permanent identification plate or markings containing manufacture date, testing, standards, maximum occupancy and weight, and equipment ID.
- Access systems to prevent accidental opening. Access doors/gates must not swing outward during the lift operation. OSHA permits outward-swinging gates for small, one-person lift platforms.
- Sufficient headroom for occupants to stand without crouching.
- In addition to hard hats, overhead protection (roof) if occupants are exposed to falling objects. Protection can't obscure visual path between the operator and the occupants.

- A suspension system designed to minimize platform movement due to movement of occupants.
- For platforms that are attachable to crane booms, the attachment mechanism must keep the platform floor within 10 degrees of level at all boom angles.

Figure 2 shows many of the design features on an actual personnel hoisting platform.

1.1.2 Platform Construction

ASME B30.23 and 29 *CFR* 1926.1431 specify minimum construction requirements for personnel lifting platforms. All welding must be in accordance with American Welding Society or ASME procedures.

Welders shall smooth all surfaces to avoid punctures or lacerations to occupants. After completion of fabrication, the manufacturer shall permanently attach an identification plate to the platform documenting the required certification testing, discussed later in this module. The platform must include a weatherproof box or placard for storing or displaying the operating instructions. Placarding shall include a full set of hand signals as described in 29 *CFR* 1926.1422, *Signals—Hand Signal Chart*.

Manufacturers shall design and assemble bridle slings or other rigging specifically for suspension of the platform equipment (*Figure 3*). The manufacturer should make these slings and attachments in such a way that it would be impractical to remove them and use the suspension system for any other purpose. It is widely accepted that the intent is to dedicate the necessary suspension system/rigging to hoisting personnel only, and not use the equipment for other tasks. Per *ASME B30.23*, slings shall have the following characteristics whenever each type is used (note that synthetic slings and ropes made from synthetic or natural fibers must not be used):

- Wire rope slings with mechanically spliced Flemish eyes must have thimbles in all eyes. Wire rope clips, wedge sockets, or knots shall not be used.
- Wire rope slings with poured socket end connections must conform to the manufacturer's or a qualified person's application instructions.
- Chain-sling suspension systems shall be made from a minimum of grade 80 chain.

All slings must use a master link or a bolt-type shackle with cotter pin for attachment to the crane hoist. The master link shall include a tag showing the rated load for the equipment and identification as a part of personnel lifting equipment so

Figure 2 Example of a suspended personnel hoisting platform.

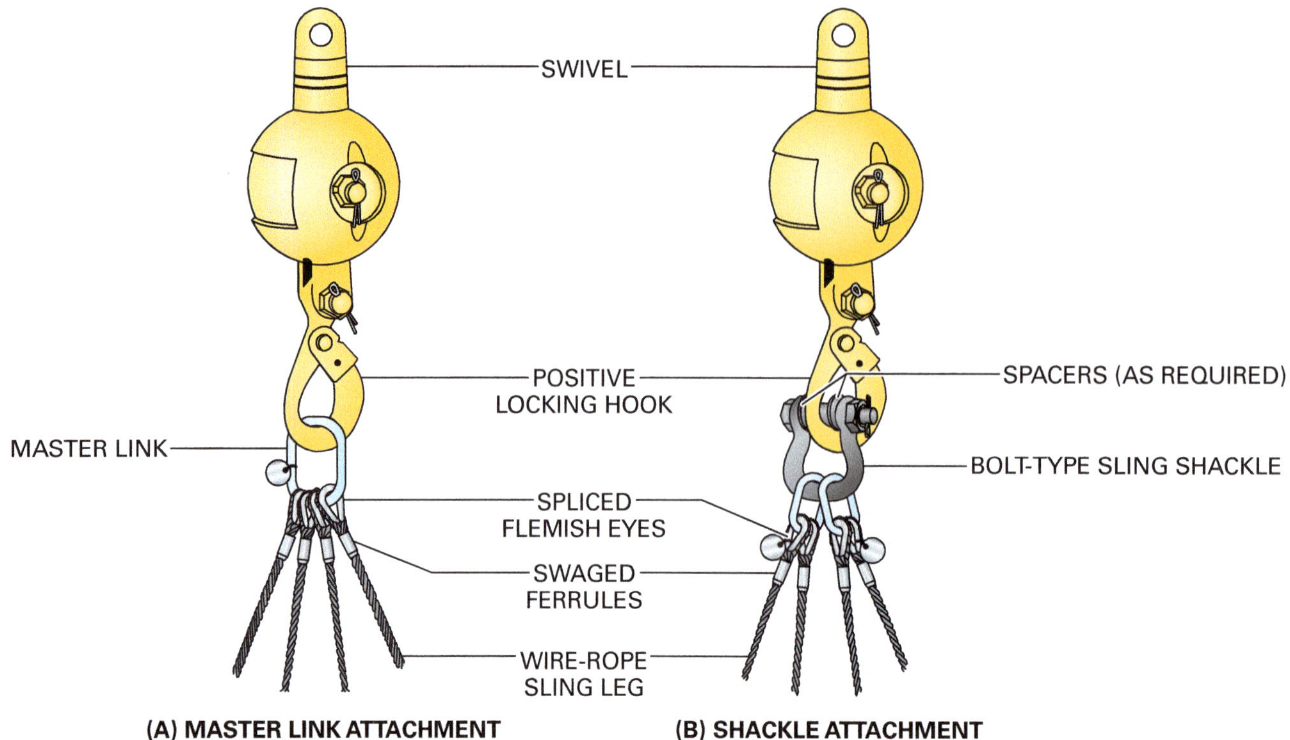

Figure 3 Hoisting rope slings fitted with spliced Flemish eyes and permanent thimbles.

it isn't used for any other purpose. *ASME B30.23* also states that rope and chain slings shall have a permanent tag or label attached to each leg that displays the rated load of the leg.

1.2.0 Testing and Inspections

The safety of lifting platform occupants is the primary purpose of the ASME and OSHA standards. However, the design and construction standards simply establish a set of expectations that manufacturers and users must meet. They can't account for construction flaws or normal wear and tear due to use. To ensure that the equipment meets and maintains the standards, builders, owners, and users must also complete certain tests and inspections.

1.2.1 Initial Manufacturing Tests

ASME B30.23 requires that after building a personnel lift platform, the manufacturer must conduct load tests of the device to ensure it meets the applicable design standards. The tests must include the load suspension attachment points, the occupant safety features, and the platform rating. Quality assurance inspectors must evaluate the attachment point welds with approved nondestructive testing (NDT).

The builder must test the platform suspension system components separately to twice (200 percent) the design load for each leg. For master-link

suspension systems, the master link must support twice the sum of the platform's weight and its design load. The manufacturer must document the satisfactory completion of all required certification testing on the platform's identification plate.

1.2.2 Proof Testing

Both *ASME B30.23* and 29 *CFR* 1926.1431 require that users conduct a special test of a lift platform at every new work site before permitting the hoisting of personnel. This proof test of the platform and its rigging entails loading the platform evenly with a weight equal to 125 percent of its rated load and suspending it for at least 5 minutes. Most platform manufacturers provide a customized test weight for each model of platform that users can easily attach and detach for this purpose (*Figure 4*).

Before and after completing the test, a competent person as defined in the OSHA standards shall inspect the platform and suspension rigging. Users must correct any defects or problems noted before using the platform for hoisting personnel. If the inspection reveals defects, users must conduct another proof test after correcting the defects.

ASME B30.23 requires that after any structural repair or modification to the platform, users must perform a proof test of 150 percent of the platform's rated load. Operators must raise a

Figure 4 Manbasket with manufacturer supplied test weight.

suspended platform to a height, then lower it at a rate exceeding 100 feet per minute (30 m/min). The operator then applies the hoist drum brake to stop the platform, and keeps it suspended for at least 5 minutes at that point. A competent person must inspect the platform. If there are any defects, they must be corrected and the platform retested before use.

1.2.3 Trial Lifts

In addition to the proof test, which tests the personnel lift hardware, 29 *CFR* 1926.1431 requires a trial lift of the platform to evaluate the lift operation itself. Users may combine a trial lift with the initial site proof test, if feasible. The OSHA standard requires that users conduct the trial lift from wherever the hoisting operation will occur, whether from ground level or from other points of origin. If the scheduled tasks require lifting occupants to different locations using a single crane setup, then a trial lift must be over the path to each location. In addition, users must complete trial lifts as follows:

- Prior to each shift during which a personnel hoist will occur
- Any time the crane moves to a new work site or returns to a previous location
- Whenever the lift route changes, unless a competent person determines the new route introduces no new factors affecting safety

Trial lifts for destinations below grade level are particularly important. Crane operators may describe this as *going negative*. There is a real possibility that a crane could run out of hoist rope if the work destination is too far below grade. Planning needs to take this possibility into account and operators must conduct such tests carefully while monitoring the remaining cable on the hoist drum.

29 *CFR* 1926.1431 also specifies other actions during the trial lift. Operators must accurately measure the load radius used for the trial lift and verify that the load will not exceed 50 percent of the crane's capacity at that radius during the lift. After the trial lift, a competent person performs the inspections discussed earlier. These must include examining the ground support, the crane, and the platform for any deficiencies. The individual must also verify that workers remove the platform's test weight.

1.2.4 Inspections and Maintenance

The ASME and OSHA standards provide only the minimum guidance regarding inspections and maintenance. Competent persons must perform inspections prior to and after every lift involving personnel, using a form similar to the *Personnel Lift Platform Pre-Lift Inspection* checklist, found in *Appendix A*. Using the checklist ensures inspection items are not skipped.

ASME B30.23 requires that both frequent and periodic platform inspections be accomplished. Frequent inspections are made at least once each day, prior to use. Periodic inspections must be conducted at least once every 12 months. Since frequent inspections are conducted often, documentation is completed at the point of use and the active checklist remains with the platform. The documents for the most recent periodic inspection must also be kept with the platform.

The platform manufacturer provides a manual for maintaining the equipment. The manual normally includes a parts list as well as guidelines for setting up a maintenance program. *ASME B30.23* requires that owners and users establish and carry out a maintenance program according to the manufacturer's instructions. Such programs will include specific inspections. Instructions also include guidelines for repairs and adjustments. Any repair or replacement of parts must meet or exceed the manufacturer's original specifications. The platform owner is responsible for maintaining records for all repairs to, or replacement of, platform structural components.

For personnel hoisting equipment adjustments and repairs, the following points apply:

* Owners and users must correct hazardous conditions identified during any inspection.
* Only certified welders may make welding repairs.
* Adjustments must only restore the correct functioning of equipment.
* Only qualified persons may repair platform rigging or suspension components.
* If any rope or chain slings show signs of wear, users must take them out of service.
* Owners and users may only make modifications approved in writing by the manufacturer.

1.3.0 Personnel Hoisting Operations

This section addresses the operational requirements and considerations that apply when hoisting personnel using mobile cranes. Due to the hazardous nature of personnel hoisting operations, 29 *CFR* 1926.1431 states that all such operations are categorized as critical lifts as defined in *ASME P30.1, Planning for Load Handling Activities.* The planning and execution of a personnel hoisting operation must be detailed, complete, and safe.

1.3.1 Key Individuals

ASME B30.23 requires that a qualified individual be appointed as personnel lift authorizer and lift director. Among the main duties of a personnel lift authorizer are the following:

* Verify the need for a personnel lift
* Ensure the equipment used meets the technical and legal requirements
* Provide final authorization for the operation to proceed
* Require the lift to follow all applicable requirements

The lift director's responsibilities include the following:

* Review and approve the lift plan
* Hold the pre-lift meeting (*Figure 5*)
* Ensure qualified persons are assigned to each position
* Ensure all administrative and legal requirements are met
* Verify that the crane operator is fit to conduct the lift and meets the rest requirements
* Designate one platform occupant as the signal person
* Supervise the lift

The crane operator is responsible for complying with the technical and operational requirements established in *ASME B30.23* pertaining to the equipment, load, and operations under the individual's direct control. The standard also places the responsibility on the operator to evaluate his or her own physical and mental competency to conduct a personnel lift. One aspect of competency involves working a limited number of hours prior to the lift operation. The ASME standard prohibits an operator from conducting the lift if the following conditions apply:

* Had less than 8 hours off prior to the current shift
* Has worked for more than 10 hours prior to beginning the lift
* Will exceed 12 hours on shift before the personnel lifting has been completed

The crane operator may not engage in any activity that will divert attention from the operation (*e.g.,* taking a call on a cell phone). The operator will respond only to the designated signal person, but must obey a Stop signal issued by anyone participating in the lift. An operator may stop a lift at any time in the interest of safety. The operator must remain at the crane's controls at all times during the lift.

The ground crew must be qualified and trained for their responsibilities, which include the following:

* Attend the pre-lift meeting
* Visually inspect the platform and its rigging for hazards prior to and during the operation
* Assist platform occupants
* Ensure the hoist line does not foul the platform
* Keep persons clear from passing underneath the platform
* Maintain continuous communications with the platform occupants and crane operator
* Must not perform any duties other than those associated with the personnel lifting operation

Figure 5 The lift director conducts the pre-lift meeting.

The platform occupants are also required to have specific training and qualifications. Their responsibilities include the following:

- Recognize hazards associated with personnel lifts
- Understand lift procedures and safety precautions
- Attend the pre-lift meeting
- Maintain platform stability
- Keep all parts of their bodies inside the platform during platform movement (occupants are permitted to extend their arms outside the platform for the purposes of positioning the platform and when serving as a signal person)
- Keep their fall protection device secured at all times unless work requires otherwise
- Know the emergency Stop hand signal
- If the platform has controls, they must be qualified to operate them

1.3.2 Fall Protection

Occupants of personnel hoist platforms shall use personal fall arrest systems (*Figure 6*), except for operations over water. For operations over water, see 29 *CFR* 1926.106, *Working Over or Near Water*. 29 *CFR* 1926.502, *Fall Protection Systems Criteria and Practices*, found in Subpart M, contains fall protection requirements applicable to crane operations. The OSHA standard outlines performance criteria as well as the design and usage of the harnesses, lanyards, and attachment hardware.

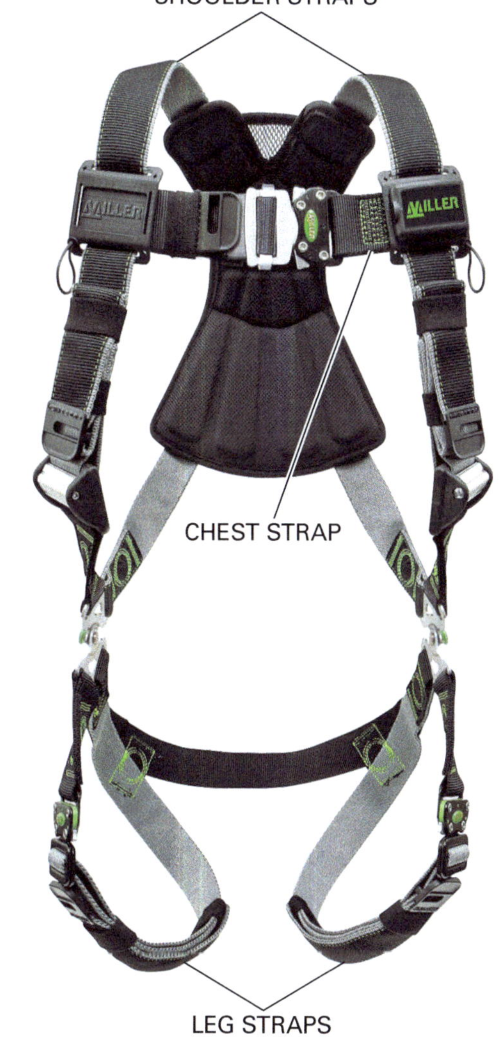

Figure 6 Fall arrest equipment.

Some of the more critical fall protection requirements include the following:

- Body belts alone are not acceptable as a personal fall arrest system.
- Harnesses must include D-rings and snap hooks that comply with OSHA and ANSI strength and fabrication specifications. Harnesses and fall arrest components should normally include labeling to this effect.
- Workers must inspect personal fall arrest systems before donning them. Any components showing wear or other defects must be removed from service.
- Unless snap hooks are of the locking type and designed for the following specific uses, platform occupants shall *not* allow snap hooks to be engaged:
  - To each other
  - Directly to webbing or wire rope
  - To D-rings retaining another snap hook
  - To any object that isn't compatible in shape to accept the snap hook
- Platform occupants shall not fasten fall arrest systems to guardrails or handrails or to any platform hoist or suspension point.
- Workers may not use fall protection equipment for any other purpose (*e.g.*, hoisting tools or materials).
- Workers shall immediately remove from service any fall arresting components subjected to impact loading while arresting a fall.

1.3.3 Crane Instrumentation and Configuration

Modern mobile cranes already include most of the hoist and instrumentation features that *ASME B30.23* and 29 *CFR* 1926.1431 require. The following list describes some of these important requirements:

- Operators must disable any feature that allows freewheeling of the hoist drum. The crane must provide the ability to positively control the lowering of the hoist line by a means other than just a friction brake.
- There must be an operable anti-two-block safety switch on the main boom or boom extension.
- A boom-angle indicator must be readily visible to the operator.
- Telescopic-boom cranes must have a boom length indicator or marks readily visible to the operator.

- Hoist hooks must have a positive-locking throat latch that prevents the lift bridle from disengaging (*Figure 7*).
- The crane's hoist controls must have a deadman feature that stops hoist rope movement when released.
- Hoist drums must have a positive method of holding their position upon the loss of hydraulics or other power source (*e.g.*, hydraulic check valves, or pinion pawls or dogs).
- There must be positive-locking devices for outriggers or stabilizers in the extended position to prevent inadvertent retraction.
- 29 *CFR* 1926.1431 permits the use of a power luffing jib for personnel hoisting, but requires that cranes have an operable jib-angle indicator readily visible to the operator, and a separate anti-two-block device for the jib hoist.

Lift planners must establish the crane configuration required for the operation. Crane assembly workers will set up the crane in the specified configuration. For any personnel hoisting operation, the operator must fully extend and lock the crane's outriggers. Critical lift plans also require an evaluation of the ground support (ground bearing pressure) at the lift site. Crane setup must include appropriate blocking, matting, or other supports as required. For land-based operations, the crane must be set up and maintained within 1 percent of level for the entire operation.

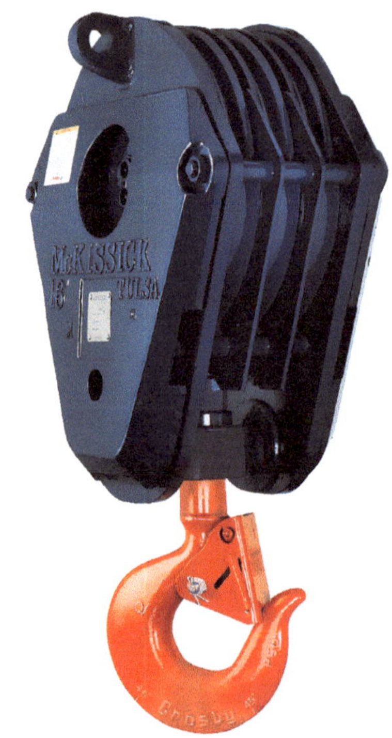

Figure 7 Crane hook with a positive-locking throat latch.

1.3.4 Personnel Hoisting Near Powerlines

Work near overhead powerlines (*Figure 8*) is especially hazardous to cranes and operators. Hoisting personnel near powerlines is an extremely dangerous operation.

The upper section of *Table 1* lists the minimum required standoff distances given in 29 *CFR* 1926.1408 and 1926.1409. OSHA allows three options for work near powerlines:

Option 1 – De-energize and ground all conductors at the work site.

Option 2 – Ensure no part of the crane, rigging, or platform approaches closer than 20 feet (for voltages under 350 kV).

Option 3 – Maintain the standoff distances listed in *Table 1* for voltages higher than 350 kV, regardless of the voltage.

In any case, OSHA prohibits any lift where the crane, rigging, or platform could encroach on the limits specified in the table. Preparations for such work should include coordination with the electrical utility to de-energize and ground the lines, if possible. Otherwise, planners must be able to accomplish the needed work from beyond the permitted standoff distances.

Figure 8 Personnel hoisted near de-energized and grounded powerlines.

Table 1 Energized Powerline Clearances for Mobile Cranes

CRANE IN OPERATION	
POWER LINE VOLTAGE IN	MINIMUM CLEARANCE IN FEET (METERS)
0 to 50	10 (3.05)
50 to 200	15 (4.60)
200 to 350	20 (6.10)
350 to 500	25 (7.62)
500 to 750	35 (10.67)
750 to 1,000	45 (13.72)
Over 1,000	Distance established by the utility owner / operator or registered professional engineer who is a qualified person in power transmission and distribution

CRANE IN TRANSIT (with no load and the boom or mast lowered) [1]	
POWER LINE (kV)	MINIMUM CLEARANCE IN FEET (METERS)
0 to 0.75	4 (1.22)
0.75 to 50	6 (1.83)
50 to 345	10 (3.05)
345 to 750	16 (4.87)
750 to 1,000	20 (6.10)
Over 1,000	Distance established by the utility owner / operator or registered professional engineer who is a qualified person in power transmission and distribution

Note 1: Environmental conditions such as fog, smoke, or precipitation may require increased clearances.

The lower section of *Table 1* lists the clearances permitted for a crane traveling with the boom and platform lowered to the travel position. OSHA allows these standoff distances only if the crane is not carrying a load. This precludes workers occupying a personnel lift platform during travel, even if otherwise permitted.

Large projects often have many contractors requiring crane services. *Appendix B* is an example of a *Power Line Awareness Permit* form. Project policies should require contractors and their crane operators to complete a form similar to *Appendix B* before performing work at the project site. Lift planners for personnel lifts near powerlines should complete the form and submit it with the lift plan for approval.

Robots Handle Live Powerlines Better than People

Safety is always a concern when working with powerlines, but the quick repair of an outage remains essential. Robotic manipulator systems are becoming more popular with major utility companies to tackle the most hazardous and critical live-line chores. An operator can remain in the safety and comfort of a control vehicle while the manipulator arms do the work. Capable of handling shocks and arcs exceeding 1,000,000 volts, the systems are immune to inclement weather or other conditions that often prevent direct intervention by human hands. Specialized equipment such as this prevents human workers in a suspended platform from being exposed to these hazards.

1.3.5 Planning and Preparations

Planning for a personnel hoisting operation must satisfy all the initial requirements described in 29 *CFR* 1926.1431 and *ASME B30.23* as summarized previously. Lift planners must also consider any other requirements established by OSHA and ASME standards relevant to critical lift operations. For example, planners must develop a written lift plan approved by the lift director. Lift planning and critical lift considerations are covered in greater detail in Module 21304 of this curriculum.

After planners develop the lift plan, the lift director reviews and approves it, then provides it to the personnel lift authorizer. The lift plan may include a lift data sheet modified as a *Personnel Platform Lift Planning and Authorization* form (similar to *Appendix C*) to help management focus on the key aspects of the personnel lift. When the personnel lift authorizer approves the plan, it passes back to the lift director for implementation.

OSHA and ASME standards require that the lift director conduct a pre-lift meeting with the lift participants to review the safety requirements, setup, and procedures to be used, as well as the work planned. Participants shall include the crane operator, lift platform occupants, signal person(s), supervisors of the planned work, and essential ground and support persons. The lift director discusses preparations, assigns responsibilities to personnel, and reviews the communications plan, including any special signals, if required. The agenda also includes general and public safety precautions, as well as emergency procedures. Participants should raise any concerns they have so that issues can be resolved before the operation begins.

Following the meeting, work crew supervisors complete the preparations according to the lift plan schedule and requirements. Prior to the lift, some of the key preparations generally include the following:

- Verify completion of all required inspections and certifications of the crane and hoisting equipment
- Perform the platform proof test, if not already completed
- Perform the applicable trial lift, if not already completed
- Establish traffic barriers/controls around the site of the lift
- Test radio communications, if used
- Verify all participants are in position and equipped with required PPE, tools, and materials
- If the lift will be near overhead powerlines, establish the conditions, barriers, and warning signs stipulated by OSHA standards
- Check the existing site weather conditions and the weather forecast
- Obtain permission to begin the lift from the lift authorizer and lift director

1.3.6 Work Practices and Crane Operations

All standard crane safety practices apply to personnel hoisting operations as well. In addition, the following list summarizes work practices that specifically apply to personnel lifts:

- The total load of the platform, occupants, rigging, and load line must not exceed 50 percent of the crane's rated capacity for the configuration and load radius.
- The standards prohibit traveling during a hoist except for cranes on fixed rails or runways, or when the employer can show that no other method exists to accomplish the work.
- The standards prohibit personnel lifts from cranes on rubber under any circumstances.
- Normally, if the platform is not resting on a stable surface for entry or egress of the occupants, it must be secured to the structure where access or the work occurs while suspended from a crane.

- If the platform is secured to any structure, the operator must not move the platform until informed that it is freely suspended.
- At the beginning of the lift, with all occupants, tools, and materials on board, operators shall conduct a test lift of a short distance (*Figure 9*). A competent person shall inspect the platform and suspension rigging to ensure that the platform is secure and properly balanced.
- Before proceeding with the lift, a competent person must verify that there are no hoist line deficiencies and the parts of line are not fouled with each other. That individual must also make sure the master link/shackle is centered above the platform and, if the hoist rope is slack, verify that all parts of line are properly seated in their sheaves.
- The ground crew must use taglines to control the platform as needed.
- The crane operator must move the platform only under the control of the designated signal person.
- The hoisting of personnel must be slow, cautious, and smooth. Raising or lowering speeds shall be less than 100 feet per minute (30 m/min) during lifts involving personnel hoisting.
- Operators may not position any part of the crane or a platform within 20 feet (6 m) of overhead powerlines up to 350 kV, or within 50 feet (15 m) of powerlines over 350 kV, unless the conditions permitted by 29 *CFR* 1926.1408 or 1926.1409 exist.
- The operator shall not hoist on any other lift lines while conducting a personnel lift.
- The operator should avoid simultaneous operation of more than one hoisting control unless doing so increases the safety of the lift operation.
- When the platform is at the work location, the operator shall set all brakes and locks on the crane hoist and slew controls before platform occupants begin working.
- During welding operations from the platform, electrode holders must prevent contact with any portion of the lift platform.
- Lift supervision must terminate hoisting operations if wind speed at the level of the platform exceeds 20 mph (32 kph); for electric storms; or for other adverse weather conditions.

Figure 9 Test lift at beginning of a personnel hoisting operation.

- For cranes on floating barges, stop personnel lifting operations if the barge list exceeds 5 degrees or the limits of the crane's load chart under list conditions. Operators also need to consider wave and wake action.
- For work over water, platform occupants may detach fall arrest devices and wear USCG-approved personal flotation devices; when over land, they must reconnect fall arrest devices. Note that the basket shown in *Figure 10* is used for transferring personnel only in marine environments under 29 *CFR* 1926.1431.
- Occupants of the platform must move slowly to avoid unbalancing the platform, and keep all body parts inside the platform during its movement. They must not intend to increase their height by sitting or standing on any part of the platform other than the floor.

Figure 10 Occupants lifted over water must wear approved flotation devices.

1.0.0   Section Review

1. For a platform equipped for a single-point lift, the attachment point must have a design factor of _____.

   a. 4
   b. 5
   c. 7
   d. 9

2. During a proof test, operators must raise a suspended platform to a height, then lower it, bringing it to a stop with the hoist drum brake. It must be lowered at a rate exceeding _____.

   a. 50 feet per minute (15 m/min)
   b. 100 feet per minute (30 m/min)
   c. 150 feet per minute (46 m/min)
   d. 200 feet per minute (61 m/min)

3. Which of the following statements about the crane operator's competency for a personnel lift is *true*?

   a. The operator must have had at least 12 hours off prior to the current shift.
   b. The operator cannot have worked for more than 4 hours prior to beginning the lift.
   c. The operator cannot exceed 12 hours on shift before the personnel lifting will be completed.
   d. The operator must have at least 1 hour prior to beginning the lift to rest outside the cab area.

SUMMARY

Consensus and safety standards prohibit the hoisting of personnel by cranes unless the employer demonstrates there is no other feasible method for the task. The design, manufacturing, and operation of personnel hoisting equipment is covered in *ASME B30.23, Personnel Lifting Systems*. 29 *CFR* 1926.1431, *Hoisting Personnel*, provides the safety regulations for these operations. According to OSHA, personnel hoisting operations are categorized as critical lifts.

Consistent testing and inspections ensure that the platform and rigging remains safe to use. Documentation of maintenance and testing helps to ensure this goal. Key personnel must plan, prepare for, and execute personnel lifts safely. Safety considerations are critical for lift plans that involve hoisting personnel near powerlines or over water.

1. The safety regulations for hoisting personnel with mobile cranes are provided in _____.

    a. 29CFR1926.1408
    b. *ASME B30.23*
    c. 29CFR1926.1431
    d. *ASME B30.5*

2. Which of the statements regarding personnel hoisting platform access doors and gates is *correct*?

    a. For larger platforms, access door and gates must swing inward.
    b. Access door and gates must be tested using a 300-lb (136 kg) weight.
    c. Access doors and gates must open without any effort.
    d. Access doors and gates must have hand- or grabrails.

3. For the purposes of personnel platform design, the ASME standard assumes an occupant weighs _____.

    a. 300 pounds
    b. 250 pounds
    c. 200 pounds
    d. 180 pounds

4. The initial builder's test of a newly constructed personnel lift platform and each leg of its bridle system must be equal to what percentage of the equipment's rated load?

    a. 100 percent
    b. 125 percent
    c. 150 percent
    d. 200 percent

5. 29 *CFR* 1926.1431 requires a proof test after any structural repair to a lifting platform. This involves lifting the platform loaded to what percentage of the platform's rated load?

    a. 200 percent
    b. 175 percent
    c. 150 percent
    d. 100 percent

6. According to 29 *CFR* 1926.1431, personnel platform trial lifts are required for all the following *except* _____.

    a. prior to each shift involving a personnel hoist
    b. when the platform occupant(s) change
    c. any time the crane moves to a new location or returns to a previous one
    d. whenever the lift route changes

7. Of the following, who is responsible for ensuring that the hoist line does not foul the personnel platform during a lift?

    a. Personnel lift authorizer
    b. Lift director
    c. Crane operator
    d. Ground crew

8. Which of the following statements pertaining to fall arrest equipment is *true*?

    a. Body belts alone are acceptable for crane hoisted personnel lifts if they are made from synthetic materials.
    b. Platform occupants may snap fall arrest harnesses to platform handrails or guardrails.
    c. Occupants can inspect and reuse a fall arrest harness again after it has arrested a fall.
    d. Platform occupants may not fasten the fall arrest harness to any platform hoist attachment point.

9. For personnel lifts near overhead powerlines, which of the following statements is *true*?

a. Standards prohibit entry of the crane boom or suspended platform within the minimum standoff distances of energized powerlines.
b. Standards permit entry into the prohibited zones near energized powerlines if the crane and platform are both grounded.
c. Standards allow personnel platforms to approach energized 480 kV powerlines to a minimum distance of 20 feet.
d. Suspended personnel platforms may not enter the OSHA-specified standoff zone near overhead powerlines under any condition.

10. Hoisting personnel with a mobile crane on rubber is _____.

a. permitted if travel with a suspended platform is required
b. permitted when the employer determines it is the only way to accomplish the task
c. never permitted
d. permitted only on a firm, stable runway

Trade Terms Introduced in This Module

Bridle slings: Lift rigging consisting of multi-leg, wire rope or chain slings.

Flemish eyes: Spliced loops formed in the end of wire rope rigging slings. For personnel hoist rigging, the eyes must also contain a thimble and be mechanically crimped with a ferrule.

Nondestructive testing (NDT): Testing the integrity of fabricated metal structures, especially welds, by various means to detect hidden flaws without destroying the object or rendering it useless. Common NDT methods include dye penetrant (PT), magnetic particle (MT), X-ray and radiography (RT), and ultrasound (UT).

Poured socket: Referring to wire rope terminations, a method of attaching a fitting by placing the end of the rope inside the fitting socket, then pouring molten zinc, another metal alloy, or a special resin into the socket. After hardening, these terminations can provide 100 percent of the wire rope's minimum breaking strength. This process is also referred to as speltering.

Proof test: A strength test of an object, usually under a load exceeding that expected under normal use, to ensure it is suitable for safe use. In personnel hoisting, a proof test requires loading the platform with a weight equal to 125 percent of its rated load and suspending it for at least 5 minutes.

Trial lift: For a personnel hoisting platform, a lift of the platform and rigging without occupants, but with an equivalent load, under the same conditions and over the same route as the lift with personnel will take.

PERSONNEL LIFT PLATFORM PRE-LIFT INSPECTION FORM

Personnel Lift Platform Pre-Lift Inspection

Date/Time: _______________________

Platform ID: _______________________

Inspector: _______________________

1. Structure	**SAT**	**UNSAT**
Load-supporting members	______	______
Load-supporting welds/fasteners	______	______
Flooring		
Barrier from toe rail to intermediate rail	______	______
Top rail	______	______
Gate hinge points and locking mechanism	______	______
Grab-/handrail	______	______
Suspension attachment points	______	______
Fall-protection anchor points	______	______

2. Attachment/Rigging Components (Circle applicable items)

	SAT	UNSAT
Pins/Eyes/Ears/Bolt-ups	______	______
Wire-rope sling/Chain/Rigid, each leg	______	______
Master link/Shackle	______	______

3. Special-Purpose Items (Overhead protection/Floatation/Platform Controls/Boom Attachment)

List:

		SAT	UNSAT
a.	_______________________	______	______
b.	_______________________	______	______
c.	_______________________	______	______

4. Markings/Documentation

	SAT	UNSAT
Platform	______	______
Placard/Manual	______	______
Suspension System	______	______

5. Comments: ____

____

____

____

____

____

____
Lift Director Date/Time

POWER LINE PLANNING FORM

POWER LINE AWARENESS PERMIT

Today's Date _________________   **Project** _________________________________

Contractor Name	
Job Address	
Telephone Number	Fax Number

Emergency Contact Number	

Survey ———————————————————————————————

Before beginning any project, you must first survey your work area to find power lines at the job site. (See job-site sketch on reverse side.)

Identify ———————————————————————————————

After finding all of the power lines at your site, identify the activities you'll be doing that may put you or your workers at risk.

Mark one or more of the following:

- ☐ Cranes (mobile or truck mounted)
- ☐ Drilling rigs
- ☐ Backhoes/excavators
- ☐ Long-handled tools
- ☐ Other tools/high-reaching equipment
- ☐ Concrete pumper

- ☐ Aerial lifts
- ☐ Dump trucks
- ☐ Ladders
- ☐ Material handling & storage
- ☐ Scaffolding
- ☐ Other _________________

Eliminate or Control ———————————————————————

After identifying the power line and high-risk activities on the job site, determine how to eliminate or control the risk of electrocution (a successful determination is often reached only after consultation with the utility).

Mark one or more of the following:

- ☐ Move the activity
- ☐ Change the activity
- ☐ Have the utility de-energize the power line
- ☐ Have the utility move the power line
- ☐ Use barrier protection (insulated sleeves)
- ☐ Use an observer

- ☐ Use warning lines with flags
- ☐ Use non-conductive tools
- ☐ Use a protective technology:
    - ☐ Insulated link
    - ☐ Boom cage guard
    - ☐ Proximity device

WARNING!

It is unlawful to operate any piece of equipment near energized power lines, within the distances specified in the table on the reverse side.

Always maintain your minimum safe clearance distance from the power line, except when the utility has de-energized and visibly grounded the power line.

Voltage (nominal, kVAC)	Minimum clearance distance (feet)
up to 50	10
over 50 to 200	15
over 200 to 350	20
over 350 to 500	25
over 500 to 750	35
over 750 to 1,000	45
over 1,000	as estimated by the utility owner/operator or registered professional engineer who is a qualified person with respect to electrical power transmission and distribution

Note: The value that follows "to" is up to and includes that value. For example, over 50 to 200 means up to and including 200kV.

*information contained in this table is from OSHA 29 CFR Sections 1926.1408 and 1926.1409 (as of 2017)

Job-Site Sketch

(Draw in location of power lines and their proximity to construction site, including such things as proposed excavations, location of heavy equipment, scaffolding, material storage areas, etc.)

Completed by ____ **Date** _________________

Title ____

Approved by ____ **Date** _________________

Title ____

PERSONNEL PLATFORM LIFT PLANNING AND AUTHORIZATION FORM

Personnel Lift Planning and Authorization   Date:

Lift Platform ID: **Lift Description:**

Project: **Units:** U.S. (ft, lb)

Crane Details   Manufacturer: Model Number:

Configuration:	Base Mount Type:	Track/Outrigger Ext: ___ ft
Boom Type:	Boom Length Used: ___ ft	Boom/Jib Angle: ___ deg.
Jib Type:	Jib Length Used: ___ ft	Tail Swing: ___ ft
Crane Ballast: ___ lb	Aux Counterweight ___ lb	
Block Capacity: ___ ton	Line Size ___ in	
Single Line Pull: ___ lb	Parts Line Used ___	Reeved Capacity: ___ lb
Heavy Lift Attachments:	Load Radius: ___ ft	Superlift Weight: ___ ton

Load Details

	Quantity	Wt. ea.	Total Weight	
Basic Weight of Platform		___ lb	___	lb
Occupants		___ lb	___	lb
Equipment/Tools		___ lb	___	lb
Materials		___ lb	___	lb
		___ lb	___	lb
		Total Weight of Loaded Platform:		lb

Rigging Data (size, type, and capacity)

	Wt. ea.	Total Weight	
	___ lb	___	lb
	___ lb	___	lb
	___ lb	___	lb
	___ lb	___	lb
	___ lb	___	lb
	___ lb	___	lb
	___ lb	___	lb
	Total Rigging Weight:		lb

Additional Weight Items

	Quantity	Wt. ea.	Total Weight	
Main Hook		___ lb	___	lb
Wire Rope		___ lb	___	lb
Other Suspended Hooks		___ lb	___	lb
Aux Boom Sheaves		___ lb	___	lb
Jib		___ lb	___	lb
		___ lb	___	lb
		___ lb	___	lb
		Total Additional Weight:		lb
		GROSS LOAD (sum of all items above):		lb

Crane Capacities   Condition:

Total Load to Crane:	___ lb	___ lb	___ lb
Planned Radius:	___ ft	___ ft	___ ft
Chart Radius Used:	___ ft	___ ft	___ ft
Chart Capacity:	___ lb	___ lb	___ lb
% of Chart Capacity:	___ %	___ %	___ %

MAXIMUM PERCENTAGE OF CHART CAPACITY ALLOWED IS   50 %

MAXIMUM PERCENTAGE OF CHART CAPACITY TO BE USED IS   ___ %

Total Suspended Load: ___ lb   Reeved Capacity: ___ lb

MAXIMUM PERCENTAGE OF REEVED CAPACITY PLANNED TO BE USED IS   ___ %

Document Attachments

○ Crane Layout	○ Lift Plan Category	○ GBP Source/Calculations
○ Rigging Hookup Diagram	○ AHA	○ Project Scope
○ Crane Load Chart	○ Risk Evaluation	○ Load Weight/CG Source
○ Foundation Details	○ Weather Forecast	○ Load/Crane Clearances
○ Crane Cribbing Arrangement	○ Load Schematic	○

Notes

	Name (Print)	Signature	Title	Date
Prepared by:				
Checked by:				
Approved:				

Answer	Section Reference	Objective
Section One		
1. c	1.1.1	1a
2. b	1.2.2	1b
3. c	1.3.1	1c

NCCER CURRICULA — USER UPDATE

NCCER makes every effort to keep its textbooks up-to-date and free of technical errors. We appreciate your help in this process. If you find an error, a typographical mistake, or an inaccuracy in NCCER's curricula, please fill out this form (or a photocopy), or complete the online form at **www.nccer.org/olf**. Be sure to include the exact module ID number, page number, a detailed description, and your recommended correction. Your input will be brought to the attention of the Authoring Team. Thank you for your assistance.

Instructors – If you have an idea for improving this textbook, or have found that additional materials were necessary to teach this module effectively, please let us know so that we may present your suggestions to the Authoring Team.

NCCER Product Development and Revision

13614 Progress Blvd., Alachua, FL 32615

Email: curriculum@nccer.org
Online: www.nccer.org/olf

❑ Trainee Guide   ❑ Lesson Plans   ❑ Exam   ❑ PowerPoints   Other ________________

Craft / Level: ____ Copyright Date: ___________

Module ID Number / Title: ____

Section Number(s): ____

Description: ____

____

____

____

Recommended Correction: ____

____

____

____

Your Name: ____

Address: ____

____

Email: ____ Phone: ___________________

Glossary

Backward stability: In crane operations, the measure of a crane steadiness in relation to the tipping axis on the opposite side of the crane from the boom.

Boom angle: In load charts, the angle between the boom relative to the horizontal plane. On telescopic-boom cranes, boom angle instruments indicate the angle between the bottom of the base section and the horizontal with the boom loaded. On lattice-boom cranes, it is the angle between the center line of the boom and a horizontal line.

Boom length: Measured from the boom foot pins to the center of the boom-point sheaves. It does not include the length of an attached jib.

Boom-point elevation: Distance from the ground to the center of the boom-point sheaves.

Bridle slings: Lift rigging consisting of multi-leg, wire rope, or chain slings.

Center of gravity (CG): The point where one can assume all of an object's mass, and therefore its weight, is concentrated. The concept is useful for determining stability, a load's balance point, and leverage.

Chicago boom: A load-hoisting device consisting of a single boom attached at its base to either a vertical post or beam in a larger structure, such as a building skeleton, or a freestanding pole stabilized by guy wires. Such booms are easily erected and suitable for repetitive hoisting operations at a fixed location.

Coefficient of friction (CF): A ratio that expresses a comparison between the force necessary to move an object over the surface of another material, and the pressure between the two materials.

Combined CG: The sum of the crane's and load's CGs at a position along an imaginary line connecting the two CGs.

Configuration: A term that addresses all aspects of a crane's condition for a given lift that can include boom assembly and position, hoisting line and accessories, base arrangement, counterweights, and other factors addressed in a load chart.

Cribbing: A stack of heavy squared timbers laid in a crisscross pattern to temporarily support loads during movement and placement.

Dynamics: The response of objects to the application of forces, moments, or torques.

Effective weight: With reference to jib/boom extensions, the weight calculated by the manufacturer which, when applied at the boom tip, has the same effect on the crane capacity as the jib and/or boom extension itself.

Flemish eyes: Spliced loops formed in the end of wire rope rigging slings. For personnel hoist rigging, the eyes must also contain a thimble and be mechanically crimped with a ferrule.

Forward stability: In crane operations, the measure of the crane's steadiness in relation to the tipping axis on the same side as the boom.

Gross capacity: A crane capacity value as displayed in the crane's load chart for a specified configuration.

Gross load: The total weight of all the items lifted by the crane. The gross capacity of the crane minus any deductions for the crane configuration results in the gross load that can be lifted. Gross load is directly related to net capacity.

Ground bearing pressure (GBP): The maximum force per unit area (pressure), measured in pounds per square foot, that can be exerted on the ground's surface without causing deformation of the surface. Can only be determined by testing.

Impact loading: Sudden forces acting on an object due to collisions or other dynamic events.

Inclined planes: Flat surfaces placed at a shallow angle that connect levels at different heights. A ramp is an example of an inclined plane.

Interpolate: To estimate or calculate an unknown intermediate value that falls between two known values.

Lift data sheet (LDS): The essential page or pages of a written lift plan that organize and summarize the important information required to develop a lift plan.

Line pull: The maximum pulling force a crane hoist drum can apply to a hoist cable, accounted for by the torque that the motor can apply to the drum and the working radius of the rope as it leaves the drum.

Load operating radius: The horizontal distance from a vertical line through the axis of crane rotation to the center of the vertical hoist line or tackle, with a load applied. Also called load radius, lifting radius, or operating radius.

Momentum: A property of a moving object that is directly proportional to both its mass and speed.

Net capacity: The crane's hoisting force available after taking into account the crane's and/or load's configurations relating to a lift. Equal to load chart gross capacity minus weight deductions.

Non-centered lift: Any lift attempted when the hoist line is not vertical or the hook is not directly above the load's center of gravity.

Nondestructive testing (NDT): Testing the integrity of fabricated metal structures, especially welds, by various means to detect hidden flaws without destroying the object or rendering it useless. Common NDT methods include dye penetrant (PT), magnetic particle (MT), X-ray and radiography (RT), and ultrasound (UT).

Offset: The angle formed by an erected boom, measured from an imaginary line extending through the center line of the main boom. The crane's load chart provides capacities for allowable jib or boom-extension offset angles.

Pendulum effect: The dynamic effects of a swinging object on the support from which it is suspended.

Poured socket: Referring to wire rope terminations, a method of attaching a fitting by placing the end of the rope inside the fitting socket, then pouring molten zinc, another metal alloy, or a special resin into the socket. After hardening, these terminations can provide 100 percent of the wire rope's minimum breaking strength. This process is also referred to as speltering.

Proof test: A strength test of an object, usually under a load exceeding that expected under normal use, to ensure it is suitable for safe use. In personnel hoisting, a proof test requires loading the platform with a weight equal to 125 percent of its rated load and suspending it for at least 5 minutes.

Rebar: Shortened form of the term *reinforcement bar*. Used to identify individual rods, bundles, and field-fabricated meshes of rods that reinforce poured concrete structures.

Ringer crane: A large, high-capacity crane erected on a circular track (the ring, usually installed directly on the ground) that supports the entire crane base.

Tipping axis: In crane operations, an imaginary horizontal line around which a crane's center of gravity would rotate as it tips.

Trial lift: For a personnel hoisting platform, a lift of the platform and rigging without occupants, but with an equivalent load, under the same conditions and over the same route as the lift with personnel will take.

Index

F

Fall protection, (21305):2, 7–8, 11, 12
Flemish eyes, (21305):1, 3, 16
Forward stability
  center of gravity (CG) and, (38301):6, 8
  defined, (38301):1, 35
  highlighted in text, (38301):5
  wind and, (38301):11–12
Friction, coefficient of (CF), (38301):24, 25–27, 35

G

GBP. *See* Ground bearing pressure (GBP)
General contractor/construction manager, responsibilities, (21304):1
Gross capacity, (21301):1, 2, 49
Gross load, (21301):1, 3, 49
Ground bearing pressure (GBP), (21304):1, 13, 20
Ground crew, qualifications and training, (21305):6
GROVE TM1500, Load Chart, (21304):23–24
Grove TM-1500 truck-mounted telescopic-boom crane
  capacity, calculating, (21301):33, 35–37
  load charts, (21301):56–57
Grove YB4409 industrial crane, capacity, calculating, (21301):32–33, 34
Guardrails, (21305):2

H

Hand rails, (21305):2
Hand signals, (21305):3
Hazards, planning for, (21304):2
Hoisting
  hoist rope, (21301):26, 28
  line pull, (21301):28
  parts of line, (21301):28–29, 30
  range diagram, (21301):29–30, 31
Hoist rope, (21301):26, 28
Hoist sheave bearings, (21301):28

I

Impact forces, (38301):13
Impact loading, (38301):1, 13, 35
Inclined planes, (38301):24, 25–27, 35
Industrial/all-purpose cranes
  capacity, calculating, (21301):32–33, 34
  configuration, (21301):9
Interpolate, (21301):9, 13, 49
ISO standards, tipping, (21301):5

J

Jib charts, (21301):25–26, 27

K

Kinetic friction, coefficient of, (38301):27

L

Lattice-boom cranes
  capacity, calculating, (21301):33, 37–38, 39–42
  configuration, (21301):10
  crawler cranes, load chart, (21301):11, 12
  deductions, (21301):4
LDS. *See* Lift data sheet (LDS)
Leverage, (21301):2
Lift data sheet (LDS)
  defined, (21304):1, 20
  HVAC condenser example
    approvals, (21304):15
    capacity calculations, (21304):10–12
    configuration, (21304):6–10, 21, 22, 23–28
    developing, (21304):5–6

    Documents and Attachments section, (21304):13, 22
    example, (21304):29
    load and rigging, (21304):10–13, 22, 23–28
    Notes section, (21304):15
    review, (21304):15
Lift director, responsibilities, (21304):2, 16–17
Lifting beams, (38301):27–28
Lift personnel
  hazards to, (21304):2
  responsibilities
    crane assembly/disassembly director, (21304):1
    crane operator, (21304):1
    crane owner, (21304):1
    crane user, (21304):1
    engineer, (21304):1
    general contractor/construction manager, (21304):1
    lift director, (21304):2
    lift planner, (21304):2
    rigger, (21304):2
    signal person, (21304):2
    site safety officer, (21304):2
    site supervisor, (21304):2
    spotter, (21304):2
    transport operator, (21304):2
Lift planner, responsibilities, (21304):2
Lift planning, (21301):29
  multiple-crane lifts, (38301):14
  spreader beams, (38301):27
Lift plans
  considerations
    capacity, (21304):4
    commercial impacts, (21304):2
    complexity, (21304):2
    environmental conditions, (21304):4
    hazards, (21304):2
    personnel responsibilities, (21304):1–2
    purposes, (21304):1
    repetitive operations, (21304):4
    standards, (21304):1
  for critical lifts, (21304):4–5, 15, 30–41
  software for, (21304):7
  for standard lifts, (21304):4–5
  standards, (21304):1
Lift plans, HVAC condenser example
  critical lifts, (21304):15
  lift data sheets
    approvals, (21304):15
    crane capacity calculations, (21304):10–12
    crane configuration, (21304):6–10, 21, 22, 23–28
    developing, (21304):5–6
    Documents and Attachments section, (21304):13, 22
    example, (21304):29
    load and rigging, (21304):10–13, 22, 23–28
    Notes section, (21304):15
    review, (21304):15
Lifts
  executing, (21304):16
  multiple-crane
    applications, (38301):13
    attachment points, (38301):14, 16–19
    planning, (38301):14
    with equalizer beam, (38301):19–22
  non-centered, (38301):1, 10–11, 35
  reviews
    post-lift, (21304):17
    pre-lift, (21304):16
  stopping, (21304):16
Line pull, (21301):25, 28, 49